Continuous Distributions in Engineering and the Applied Sciences – Part II

Synthesis Lectures on Mathematics and Statistics

Editor
Steven G. Krantz, *Washington University, St. Louis*

An Introduction to Proofs with Set Theory
Daniel Ashlock and Colin Lee
2020

Discrete Distributions in Engineering and the Applied Sciences
Rajan Chattamvelli and Ramalingam Shanmugam
2020

Affine Arithmetic Based Solution of Uncertain Static and Dynamic Problems
Snehashish Chakraverty and Saudamini Rout
2020

Time-Fractional Order Biological Systems with Uncertain Parameters
Snehashish Chakraverty, Rajarama Mohan Jena, and Subrat Kumar Jena
2020

Fast Start Advanced Calculus
Daniel Ashlock
2019

Fast Start Integral Calculus
Daniel Ashlock
2019

Fast Start Differential Calculus
Daniel Ashlock
2019

Introduction to Statistics Using R
Mustapha Akinkunmi
2019

Inverse Obstacle Scattering with Non-Over-Determined Scattering Data
Alexander G. Ramm
2019

Analytical Techniques for Solving Nonlinear Partial Differential Equations
Daniel J. Arrigo
2019

Aspects of Differential Geometry IV
Esteban Calviño-Louzao, Eduardo García-Río, Peter Gilkey, JeongHyeong Park, and Ramón Vázquez-Lorenzo
2019

Symmetry Problems. The Navier–Stokes Problem.
Alexander G. Ramm
2019

Continuous Distributions in Engineering and the Applied Sciences – Part II
Rajan Chattamvelli and Ramalingam Shanmugam

ISBN: 978-3-031-01307-2   paperback
ISBN: 978-3-031-02435-1   ebook
ISBN: 978-3-031-00281-6   hardcover

DOI 10.1007/978-3-031-02435-1

A Publication in the Springer series
SYNTHESIS LECTURES ON MATHEMATICS AND STATISTICS

Lecture #43
Series Editor: Steven G. Krantz, *Washington University, St. Louis*
Series ISSN
Print 1938-1743   Electronic 1938-1751

Continuous Distributions in Engineering and the Applied Sciences – Part II

Rajan Chattamvelli
VIT University, Vellore, Tamil Nadu

Ramalingam Shanmugam
Texas State University, San Marcos, Texas

SYNTHESIS LECTURES ON MATHEMATICS AND STATISTICS #43

ABSTRACT

This is the second part of our book on continuous statistical distributions. It covers inverse-Gaussian, Birnbaum-Saunders, Pareto, Laplace, central χ^2, T, F, Weibull, Rayleigh, Maxwell, and extreme value distributions. Important properties of these distribution are documented, and most common practical applications are discussed. This book can be used as a reference material for graduate courses in engineering statistics, mathematical statistics, and econometrics. Professionals and practitioners working in various fields will also find some of the chapters to be useful.

Although an extensive literature exists on each of these distributions, we were forced to limit the size of each chapter and the number of references given at the end due to the publishing plan of this book that limits its size. Nevertheless, we gratefully acknowledge the contribution of all those authors whose names have been left out.

Some knowledge in introductory algebra and college calculus is assumed throughout the book. Integration is extensively used in several chapters, and many results discussed in Part I (Chapters 1 to 9) of our book are used in this volume.

Chapter 10 is on Inverse Gaussian distribution and its extensions. The Birnbaum-Saunders distribution and its extensions along with applications in actuarial sciences is discussed in Chapter 11. Chapter 12 discusses Pareto distribution and its extensions. The Laplace distribution and its applications in navigational errors is discussed in the next chapter. This is followed by central chi-squared distribution and its applications in statistical inference, bioinformatics and genomics. Chapter 15 discusses Student's T distribution, its extensions and applications in statistical inference. The F distribution and its applications in statistical inference appears next. Chapter 17 is on Weibull distribution and its applications in geology and reliability engineering. Next two chapters are on Rayleigh and Maxwell distributions and its applications in communications, wind energy modeling, kinetic gas theory, nuclear and thermal engineering, and physical chemistry. The last chapter is on Gumbel distribution, its applications in the law of rare exceedances.

Suggestions for improvement are welcome. Please send them to rajan.chattamvelli@vit.ac.in.

KEYWORDS

Birnbaum–Saunders distribution, chi-squared distribution, electronics, Gumbel distribution, Laplace distribution, Maxwell distribution, mean deviation, Pareto distribution, Rayleigh distribution, size-biased distributions, survival function, truncated distributions, Weibull distribution

Contents

List of Figures

List of Tables

Preface

Continuous distributions are encountered in several engineering fields. They are used either to model a continuous variate (like time, temperature, pressure, amount of rainfall) or to approximate a discrete variate. This book (in two parts) introduces the most common continuous univariate distributions. A comprehensive treatment requires entire volumes by itself, as the literature on these distributions are extensive and ever-increasing. Hence, only the most important results that are of practical interest to engineers and researchers in various fields are included. Professionals working in some fields usually encounter only a few of the distributions for which probability mass function (PMF), probability density function (PDF), cumulative distribution function (CDF), survival function (SF) (complement of CDF), or hazard rate are needed for special variable values. These appear in respective chapters.

Rajan Chattamvelli and Ramalingam Shanmugam
July 2021

Glossary of Terms

Term	Meaning
ALD	Asymmetric Laplace Distribution
ANOR	Analysis Of Reciprocals
ANOVA	Analysis Of Variance
BSD	Birnbaum–Saunders Distribution
BSR	Birnbaum–Saunders Regression
ChF	Characteristic Function
CFAR	Constant False Alarm Rate
CLD	Classical Laplace Distribution
CST	Classical Student's T distribution
CT	Contingency Table
CV	Coefficient of Variation
DFR	Decreasing Failure Rate
DLD	Double Lomax Distribution
DoF	Degrees of Freedom
DSP	Digital Signal Processing
EVD	Extreme-Value Distribution
FSTD	Folded Student's T Distribution
GF	Generating Function
GPD	Generalized Pareto Distribution
HF	Hazard Function
IBF	Incomplete Beta Function
IDF	Inverse Distribution Function
IFR	Increasing Failure Rate
IGD	Inverse Gaussian Distribution
IID	Independently and Identically Distributed
INID	Independent, Not Identically Distributed
IRD	Inverse Rayleigh Distribution
IWD	Inverse Weibull Distribution
LBM	Linear Brownian motion
LBS	Log-Birnbaum–Saunders Distribution
LDL	Lower Detection Limit
LLD	Log-Laplace Distribution

LND	Log-Normal Distribution
LRT	Likelihood Ratio Tests
LST	Log-Student's T Distribution
MBD	Maxwell–Boltzman Distribution
MD	Mean Deviation
MGF	Moment Generating Function
MLE	Maximum Likelihood Estimate
MoM	Method of Moments Estimate
MRI	Magnetic Resonance Imaging
MRL	Mean Residual Life
MVUE	Minimum Variance Unbiased Estimate
NCT	Noncentral T distribution
PDF	Probability Density Function
PFR	Percentage Failure Rate
PMD	Population Mahalanobis Distance
PMF	Probability Mass Function
RFM	Rayleigh Fading Model
RIG	Reciprocal Inverse Gaussian Distribution
SED	Standard Exponential Distribution
SF	Survival Function
SGD	Standard Gumbel Distribution
SkLD	Skew Laplace Distribution
SLD	Standard Laplace Distribution
SNR	Signal to Noise Ratio
SPRT	Sequential Probability Ratio Tests
SRD	Standard Rayleigh Distribution
STF	Student T Filters
TTF	Time To Failure
UAV	Unmanned Aerial Vehicles
UDL	Upper Detection Limit
WPP	Weibull Probability Plot

CHAPTER 10

Inverse Gaussian Distribution

10.1 INTRODUCTION

The Inverse Gaussian Distribution (IGD) first appeared in the works of Louis Bachelier (1900) in his doctoral thesis on the theory of speculation, and subsequently in Schrödinger (1915) in the theory of linear Brownian motion (LBM). The probability density function (PDF[1]) of two-parameter version is

$$f(x; \mu, \lambda) = \sqrt{\lambda/(2\pi x^3)} \, \exp\{-\frac{\lambda}{2\mu^2 x}(x-\mu)^2\}, \text{ for } x > 0. \tag{10.1}$$

The name "inverse Gaussian" was suggested by Tweedie (1957) [219], who noticed that there exists an inverse relation between the cumulant generating function (CGF) of the distribution of time needed to cover a constant distance, and the distance covered during constant time in LBM.[2] This is also called Tweedie's distribution, Wald's distribution, first-passage-time distribution of Brownian motion, and as Hadwiger function in reproduction modeling (biological sciences).

Wald (1947) [225] derived the IGD as a limiting distribution of sample size in sequential probability ratio tests (SPRT). As shown later, Wald's distribution is a special case of IGD. It is denoted by $IG(\mu, \lambda)$ or $IGD(\mu, \lambda)$, where μ is the location parameter and λ is the shape parameter. An alternate notation is $N^{-1}(\mu, \lambda)$ where N denotes the normal (Gaussian) law. Both the parameters are positive (because λ that appears inside the radical must be positive, and the variance being $\mu^3/\lambda > 0$ implies that $\mu > 0$).

10.1.1 ALTERNATE REPRESENTATIONS

As $1/\sqrt{2\pi} = 0.39894228$, the PDF can be written as

$$f(x; \mu, \lambda) = 0.39894228 \, (\lambda/x^3)^{1/2} \, \exp\{-\frac{\lambda}{2\mu^2 x}(x-\mu)^2\}, \text{ for } x > 0. \tag{10.2}$$

The μ is denoted as a, and λ by σ in Brownian motion and some other fields. Separate the constant in (10.1) to get another form as

$$f(x; \mu, \lambda) = \sqrt{\lambda/(2\pi)} \, x^{-3/2} \, \exp\{-\frac{\lambda}{2\mu^2}(x-\mu)^2/x\}, \text{ for } x > 0. \tag{10.3}$$

---

[1]PDF is called electron probability density function or electron density in thermodynamics and computational chemistry.
[2]Tweedie (1947) [218] also found such a relation among the binomial and negative binomial distributions, and between Poisson and gamma distributions.

Expand $(x - \mu)^2$ as a quadratic and divide each term by μx to get

$$f(x; \mu, \lambda) = \sqrt{\lambda/(2\pi x^3)} \, \exp\{-\frac{\lambda}{2\mu}(x/\mu - 2 + \mu/x)\}, \text{ for } x > 0, \mu, \lambda > 0. \quad (10.4)$$

Take the constant in the exponent as a separate multiplier and put $\lambda/\mu = \delta$ to get another form

$$f(x; \mu, \delta) = \sqrt{\delta\mu/(2\pi x^3)} \, \exp(\delta) \, \exp\{-\frac{\delta}{2}(x/\mu + \mu/x)\}, \text{ for } x > 0, \mu, \delta > 0. \quad (10.5)$$

Taking $\phi = \lambda/\mu^2$ results in

$$f(x; \lambda, \phi) = \sqrt{\lambda/(2\pi)} x^{-3/2} \, \exp(\lambda\phi/2) \, \exp\{-\frac{1}{2}(\lambda x - 1 + \phi x)\}, \text{ for } x > 0. \quad (10.6)$$

Put $\lambda = t^2$ and $1/\mu = \tau^2$ to get the alternate representation

$$f(x; t, \tau) = t/\sqrt{2\pi x^3} \, \exp(t^2\tau^2) \, \exp\{-t^2/(2x) - \tau^2 x/2)\}, \text{ for } x > 0, t, \tau > 0. \quad (10.7)$$

Tweedie discussed three other representations using $\lambda = \phi\mu$

$$f_1(x; \alpha, \lambda) = \sqrt{\lambda/(2\pi x^3)} \, \exp(\lambda\sqrt{2\alpha}) \, \exp\{-\lambda/(2x) - \alpha\lambda x\}, \quad (10.8)$$

$$f_2(x; \mu, \phi) = \sqrt{(\mu\phi)/2\pi x^3} \, \exp(\phi) \, \exp\{-\mu\phi/(2x) - \phi x/(2\mu))\}, \quad (10.9)$$

and

$$f_3(x; \phi, \lambda) = \sqrt{\lambda/2\pi x^3} \, \exp(\phi) \, \exp\{-\lambda/(2x) - \phi^2 x/(2\lambda))\}. \quad (10.10)$$

These forms are used in different fields, and are related as $f(x; \mu, \lambda) = f_2(x/\mu; 1, \phi)/\mu = f_3(x/\lambda; \phi, 1)/\lambda$. As shown below, the mean is μ and the variance is μ^3/λ. Set $\mu^3/\lambda = \mu$ so that the mean and variance are equal, and $\mu^2 = \lambda$. This results in a one-parameter IGD(μ) with PDF

$$f(x; \mu) = \mu/\sqrt{2\pi x^3} \, \exp(-(x - \mu)^2/(2x)), \text{ for } x > 0. \quad (10.11)$$

Another one-parameter distribution results when $\mu \to \infty$ in (10.1) with PDF

$$f(x; \lambda) = \sqrt{\lambda/(2\pi x^3)} \, \exp(-\lambda/(2x)), \text{ for } \quad x > 0, \lambda > 0. \quad (10.12)$$

Ahmed et al. (2008) introduced a re-parameterization with PDF

$$f(x; \lambda, \mu) = \lambda/(\mu\sqrt{2\pi})(\mu/x)^{3/2} \, \exp\left(-\frac{1}{2}(\sqrt{x/\mu} - \lambda\sqrt{\mu/x})^2\right), \text{ for } \quad x > 0, \lambda > 0.$$

$$(10.13)$$

10.2 RELATION TO OTHER DISTRIBUTIONS

It is called standard Wald's distribution when $\mu = 1$. The IGD tends to a normal (Gaussian) distribution as $\lambda \to \infty$. It is related to the standard normal PDF using the transformation $Y = \sqrt{\lambda/X}(X/\mu - 1)$. The PDF of Y then becomes (Chhikara and Folks (1989) [49], Seshadri (1999)) [198]

$$g(y; \mu, \lambda) = \left(1 - y/\sqrt{4\lambda/\mu + y^2}\right)\phi(y), \quad -\infty < y < \infty. \tag{10.14}$$

As λ/μ is invariant under a scale-transformation, the following relationships hold for the PDF

$$f(x; \mu, \lambda) = (1/\mu)f(x/\mu; 1, \lambda/\mu) = (1/\lambda)f(x/\lambda; \lambda/\mu, 1), \quad 0 < x < \infty. \tag{10.15}$$

If $X \sim \text{IGD}(\mu, \lambda)$, then $Y = \lambda(X - \mu)^2/(\mu^2 X)$ has (χ_1^2) distribution (Chapter 14). If λ is held constant and $\mu \to \infty$, $\text{IGD}(\mu, \lambda)$ approaches a gamma distribution $\text{GAMMA}(\lambda/2, 1/2)$. The transformation $Z = \lambda X/\mu^2$ is considered to be a standard form $\text{IG}(\phi, \phi^2)$ where $\phi = \lambda/\mu$ is assumed to be a single parameter, which is used for tabulation. Under this transformation $E(Z)$ and $Var(Z)$ are both equal to ϕ. The ratio $1/\phi = \delta = \mu/\lambda$ is the square of the coefficient of variation (CV). Takagi et al. (1997) [211] proved that the IGD and lognormal distribution (LND) agree well when the CV is less than one, and that the IGD is sharper when $\text{CV} > \sqrt{2}$ (See Figure 10.1b).

Writing the exponent as $(\frac{x-\mu}{\mu})^2 = (\frac{x}{\mu} - 1)^2$ and letting $\mu \to \infty$ this becomes $(-1)^2 = 1$. The resulting distribution is called one-parameter IGD:

$$f(x; \lambda) = \sqrt{\lambda/2\pi x^3} \, \exp\{-\lambda/(2x)\} = 0.3989422804 \, \sqrt{\lambda/x^3} \, \exp\{-\lambda/(2x)\}, \tag{10.16}$$

which is the distribution of first passage time of drift-free Brownian motion. If the diffusion constant is D and drift velocity is v, the time taken to reach a point at distance x from the starting position (origin) for the first time is $IG(x/v, x^2/(2D))$.

Another such form was obtained by Iliescu and Vodâ (1981) [96] by taking $\lambda = \mu^2$ as $\text{IG}(\mu, \mu^2)$ with PDF

$$f(x; \mu) = \mu/\sqrt{2\pi x^3} \, \exp\{-\frac{1}{2}(x - \mu)^2/x\} = 0.39894 \, (\mu/\sqrt{x^3}) \, \exp\{-\frac{1}{2}(x - \mu)^2/x\}. \tag{10.17}$$

The reciprocal IG distribution (RIG) (also called random-walk distribution), which is the distribution of $Y = 1/X$ has PDF

$$f(y; \mu, \lambda) = \sqrt{\lambda/(2\pi y^3)} \, \exp\{\delta\} \, \exp\{-\lambda(y/2 - 1/(2\mu^2 y))\}, \quad 0 < y < \infty \tag{10.18}$$

with cumulants $\kappa_1 = 1/\lambda + 1/\mu = (1/\lambda) \, (1 + \phi) = (1/\mu) \, (1 + \delta)$, and $\kappa_2 = 1/(\mu\lambda) + 2/\lambda^2 = (1/\lambda) \, (1/\mu + 2/\lambda) = (1/(\lambda\mu))(1 + 2\delta)$. The mode of RIG is at $(1/\mu) \left(\sqrt{1 + \mu^2/(4\lambda^2)} - \mu/(2\lambda)\right) = (1/\mu) \left(\sqrt{1 + \delta^2/4} - \delta/2\right)$.

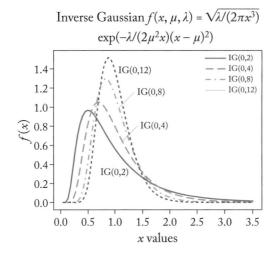

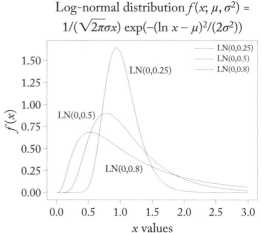

Figure 10.1: Inverse Gaussian distributions.

10.3   PROPERTIES OF IGD

This distribution is unimodal and positively skewed (Figure 10.1). If $X \sim IG(\mu, \lambda)$, then $cX \sim IG(c\mu, c\lambda)$ where $c > 0$ is a constant. The IGD obeys a reproductive property. If $X_1 \sim IG(\lambda_1, \mu)$ and $X_2 \sim IG(\lambda_2, \mu)$ are independently and identically distributed (IID), then $X_1 + X_2 \sim IG(\lambda_1 + \lambda_2, \mu)$. Such a relationship also holds among RIG distributions. In particular, the mean $\overline{X}$ of n IID IG distributed variates is also distributed as $IG(\mu, n\lambda)$. Thus, the MLE (Maximum Likelihood Estimate) of μ is the sample mean. The distribution of the MLE of the reciprocal $1/\lambda$ is χ^2 type. In addition, $\overline{X}$ and $V = \sum_k (1/X_k - 1/\overline{X})$ are independently distributed, and $\lambda \sum_k (1/X_k - 1/\overline{X})$ is distributed as χ^2_{n-1}, which is the same as the distribution of $n \, s^2/\sigma^2$ used in the analysis of variance (ANOVA) under normality assumption.

Whereas the standard decomposition of sum of squares as $\sum_{k=1}^{n}(x_k - \mu)^2 = \sum_{k=1}^{n}(x_k - \overline{x})^2 + n(\overline{x} - \mu)^2$ is used in ANOVA and various tests of hypotheses under normality assumption, the analogous IG decomposition is $(1/\mu^2)\sum_{k=1}^{n}(x_k - \mu)^2/x_k = \sum_{k=1}^{n}(1/x_k - 1/\overline{x}) + (n/\mu^2)(\overline{x} - \mu)^2/\overline{x}$. The transformation $Y = X/\mu$ results in a distribution with mode approaching one when $\delta \to \infty$, whereas $Y = X/\lambda$ results in the mode of Y to approach $1/3$ as $\delta \to 0$. Linear combinations of IGD are IG distributed. In particular, if X_i's are IGD(μ_i, λ_i) then $\sum_{i=1}^{n} \lambda_i/\mu_i^2 X_i$ is IGD$(\sum_{i=1}^{n} \lambda_i/\mu_i, (\sum_{i=1}^{n} \lambda_i/\mu_i)^2)$.

10.3.1   MOMENTS AND GENERATING FUNCTIONS

The mean is μ, but the mode depends on both μ, λ as $\mu\left((1 + 9/(4\delta^2))^{1/2} - 3/(2\delta)\right)$ where $\delta = \lambda/\mu$. For $\delta = 1$ the mode is 0.3027756 times the mean, whereas for $\delta = 2$ it is half the mean. In fact, as the second factor $3/(2\delta)$ is less than the first one, the mode is less than the mean. As δ becomes large, the mode slowly converges to the mean from below (it is $0.744\,\mu$ for $\delta = 5$, 0.861187μ for $\delta = 10$, 0.985028μ for $\delta = 100$), which shows that the convergence is slow.

Multiply (10.3) by e^{itx} and integrate over the range to get

$$\phi(t) = \exp\left(\frac{\lambda}{\mu}\{1 - (1 - (2i\mu^2 t/\lambda)^{1/2})\}\right) = \exp\left(\delta\{1 - (1 - (2i\mu t/\delta)^{1/2})\}\right). \tag{10.19}$$

This can be represented in terms of erfc() as

$$\phi(t) = \frac{1}{2}\left(\mathrm{erfc}(\sqrt{\lambda/x}(\mu - x)/\mu\sqrt{2}) + \theta^{2\lambda/\mu}\mathrm{erfc}(\sqrt{\lambda/x}(\mu - x)/\mu\sqrt{2})\right), \tag{10.20}$$

where $\theta = 1$ for standard IGD.

Problem 10.1   Prove that the mode as a function of δ is least when $\delta = 2$ for fixed μ, and occurs at $x = \mu/2$ with modal value $2\mu\sqrt{2/(\pi e)} = 0.967882898\mu$.

Problem 10.2   If $x_1, x_2, \ldots, x_n$ is a random sample of size n from an IG(μ, λ) distribution, prove that $E(1/\bar{x}_n) = \frac{1}{\mu}(1 + \mu/(n\lambda))$.

The variance also depends on both μ, λ as $\sigma^2 = \mu^3/\lambda = \mu^2/\delta$. The ratio $\sigma^2/\mu = \mu^2/\lambda = \mu/\delta$, showing that the variance is greater than the mean when $\mu > \delta$.

The kth raw moment (about the origin) is given by $E(X^k) = \mu^k \sum_{j=0}^{k-1}(k - 1 + j)!/[j!(k - 1 - j)!](\mu/(2\lambda))^j$. From this the first four raw moments follow as $\mu_1' = \mu$, $\mu_2' = \mu^2[1 + \mu/\lambda]$, $\mu_3' = \mu^3[1 + 3\phi(1 + \phi)]$, $\mu_4' = \mu^4[1 + 3\phi(2 + 5\phi(1 + \phi))]$ where $\phi = \mu/\lambda$. The central moments are $\mu_2 = \mu^2\phi$, $\mu_3 = 3\mu^3\phi^2$, $\mu_4 = 3\mu^4\phi^2(1 + 5\phi)$. The first inverse moment is $E(1/X) = 1/\mu + 1/\lambda$, which tends to zero when both μ and λ becomes large. The CGF of IG(μ, λ) is $K_x(t, \mu, \lambda) = (\lambda/\mu)\left(1 - (1 - 2\mu^2 t/\lambda)^{1/2}\right)$ for $t < \lambda/(2\mu^2)$, which in terms of $\delta = (\lambda/\mu)$ is $\delta\left(1 - (1 - 2\mu t/\delta)^{1/2}\right)$. From this the cumulants follow as $\kappa_1 = \mu$, $\kappa_r = 1.3.5.\ldots.(2r - 3)\,\lambda^{1-r}\,\mu^{1-2r} = 1.3.5.\ldots.(2r - 3)\,\delta^{1-r}\,\mu^r$ for $r \geq 2$.

The coefficient of skewness is $3\sqrt{\mu/\lambda}$, so that an analyst can choose between a variety of distributional shapes. It has greater skewness and sharper peak than the gamma distribution. The skewness decreases as λ becomes large relative to μ. The skewness can be increased without limit by letting $\mu \to \infty$ by keeping λ fixed.

The mean deviation (MD) is given by

$$\mathrm{MD} = 2\int_0^\mu \left\{\Phi\left(\sqrt{\lambda/x}\,[x/\mu - 1]\right) + e^{2\lambda/\mu}\Phi\left(-\sqrt{\lambda/x}\,[x/\mu + 1]\right)\right\} dx. \tag{10.21}$$

This can be simplified as $4\exp(2\lambda/\mu)\,\Phi(-2\sqrt{\lambda/\mu})$. The ratio MD/SD is $\sqrt{2/\pi}(1 - \mu/(4\lambda))$. This approaches 0.5984 when $\lambda \to \mu$. The kurtosis is $3 + 15(\mu/\lambda)$, showing that it is always leptokurtic. The excess kurtosis tends to zero as $\lambda \to \infty$ for fixed μ.

The MLE estimates are found from the following log-likelihood function

$$\ln(L(x_1,\ldots,x_n,\mu,\lambda)) = \frac{n}{2}\ln(\lambda) - \frac{n}{2}\ln(2\pi) - \frac{3}{2}\sum_{k=1}^{n}\ln(x_k)\sum_{k=1}^{n}\lambda(x_k-\mu)^2/(2\mu^2 x_k),$$

(10.22)

from which $\hat{\mu} = \bar{x}_n$ and $1/\hat{\lambda} = \frac{1}{n}\sum_k(1/x_k - 1/\hat{\mu}) = n/V$ (whereas the minimum variance unbiased estimate (MVUE) is $(n-3)/V$ for $n > 3$, where $V = \sum_k(1/X_k - 1/\bar{X})$). Yoon, Kim, and Song (2020) [236] use adjusted MLE after comparing various parameter estimation methods.

The moments and inverse moments are related as $E(X/\mu)^{-r} = E(X/\mu)^{r+1}$, or equivalently $E(X^{r+1}) = \mu^{2r+1}E(X^{-r})$ where μ is the mean, and negative index values denote inverse moments. In particular, $E(X^2) = \mu^3 E(X^{-1})$. Thus, all negative and positive moments exist for the IG and RIG distributions. Mudholker and Natarajan (2002) [156] called this the IG-symmetry, which is also satisfied by the log-normal distribution [47] for which $E(X/\mu)^{-r} = \exp(r(r+1)\sigma^2/2) = E(X/\mu)^{r+1}$, as well as few others. These belong to the standard family of distributions of the form

$$f(x,c) = 1/(x\sqrt{2\pi})\exp\{-(\log(x))^2/2\}\times(1 + c\sin(2\pi\log(x)))\quad\text{for }|c| < 1\quad(10.23)$$

which reduces to standard log-normal law for $c = 0$. Mudholker and Natarajan (2002) also extended the concept to fractional moments by assuming that r can be any finite real number, and observed that scale mixtures of IG-distributed independent random variables with mean μ are IG-symmetric about the same mean. See Gong, Lee, and Kang (2020) [80] for generalized MGF and its applications to wireless communications. See Table 10.1 for further properties.

10.3.2   TAIL AREAS

The CDF is expressible in terms of the standard normal CDF as

$$F(x;\mu,\lambda) = \Phi\left(\sqrt{\frac{\lambda}{x}}\left[\frac{x}{\mu}-1\right]\right) + e^{2\lambda/\mu}\Phi\left(-\sqrt{\frac{\lambda}{x}}\left[\frac{x}{\mu}+1\right]\right).\qquad(10.24)$$

When $\mu = 1$, the CDF can be expressed in terms of standard normal CDF as

$$F_x(\lambda) = \Phi((x-1)\sqrt{\lambda/x}) + e^{2\lambda}\;\Phi(-(x+1)\sqrt{\lambda/x}).\qquad(10.25)$$

See [49] for approximations.

10.3.3   EXTENSIONS OF IG DISTRIBUTION

Size-biased distributions arise in various applied fields, as when observations are recorded with probability proportional to size (PPS). They are special case of weighted distributions of the form $g(x;\theta) = w(x)f(x;\theta)/E(w(x))$ where $E(w(x)) < \infty$ with weight function $w(x) = x$. A weighted IGD (WIGD) with weight $w(x) = x^2\exp(\lambda/(2x))$ is used in survival analysis. Size-biased distribution of order α results when the weight is chosen as $w(x) = x^\alpha$, which finds applications in ecology and forestry. A size-biased IG (SBIG) distribution can be obtained from the

Table 10.1: Properties of IGD $f(x; \mu, \lambda) = \sqrt{\lambda/(2\pi x^3)} \, \exp\{-\frac{\lambda}{2\mu^2 x}(x-\mu)^2\}$

Property	Expression	Comments
Range of X	$x \geq 0$	Continuous
Mean	μ	
Median	$\log(2)/\lambda$	
Mode	$\mu((1 + 9/(4\delta^2))^{1/2} - 3/(2\delta))$	
Variance	$\sigma^2 = \mu^3/\lambda = \mu^2/\delta$	$\delta = \lambda/\mu$
Skewness	$\gamma_1 = 3/\sqrt{\delta}$	
Kurtosis	$\beta_2 = 3 + 15/\delta$	Leptokurtic
Mean dev.	$E\|X - \mu\| = 4 \exp(2\lambda/\mu)\Phi(-2\sqrt{\lambda/\mu})$	
CV	$1/\sqrt{\delta}$	$\delta = \lambda/\mu$
CDF	$\Phi\left(\sqrt{\frac{\lambda}{x}}\left[\frac{x}{\mu} - 1\right]\right) e^{2\lambda/\mu} \, \Phi\left(-\sqrt{\frac{\lambda}{x}}\left[\frac{x}{\mu} + 1\right]\right)$	
Cumulants	$\kappa_r = 1.3.5 \cdots (2r - 3)\mu^{2r-1}/\lambda^{r-1}$	$r \geq 2$
MGF	$\exp(\delta(1 - (1 - 2\mu^2 t/\lambda)^{1/2}))$	
CGF	$\delta(1 - [1 + 2\mu^2 it/\lambda]^{1/2})$	
ChF	$\exp(\delta(1 - (1 - 2\mu^2 it/\lambda)^{1/2}))$	
Additivity	$X_i \sim \text{IG}(\mu, \lambda) \Rightarrow \sum_i X_i \sim \text{IG}(n\mu, n^2\lambda)$ $X_i \sim \text{IG}(\mu_i, \lambda_i^2) \Rightarrow \sum_i X_i \sim \text{IG}(\sum_i \mu, 2(\sum_i \mu^2))$ $X_i \sim \text{IG}(\mu, \lambda) \Rightarrow \bar{x} \sim \text{IG}(\mu, n\lambda)$	

Approaches normality as $\lambda \to \infty$, otherwise it is skewed. The cumulant generating function of $\text{IG}(\mu, \lambda)$ is the inverse of the CGF of normal distribution.

relationship $\mu = \int_0^\infty x f(x; \mu, \lambda) dx$ by dividing both sides by μ to get the PDF

$$f(x; \mu, \lambda) = \sqrt{\lambda/(2\pi\mu^2 x)} \, \exp\{-\frac{\lambda}{2\mu^2 x}(x-\mu)^2\}, \text{ for } x > 0. \qquad (10.26)$$

If $X \sim \text{IG}(\mu, \lambda)$ then $Y = \mu^2/X \sim \text{SBIG}(\mu, \lambda)$ and vice versa. It finds applications in reliability, survival analysis, and medical sciences. A "split IG distribution" can be defined when the behavior of a process is different on either side of μ by splitting the frequency for $x \leq \mu$ and $x > \mu$:

$$f(y; \mu, \lambda, c) = \sqrt{\lambda/(\pi y^3(1+c)^2)} \exp(\delta)$$

$$\begin{cases} \exp\{-\lambda(y/2 - 1/(2\mu^2 y))\} & \text{for} \quad 0 < y \leq \mu; \\ \exp\{-\lambda/c^2(y/2 - 1/(2\mu^2 y))\} & \text{for} \quad \mu > y; \\ 0 & \text{otherwise}, \end{cases} \quad (10.27)$$

where c is the imbalance parameter. A mode-based reparametrized IG (rIG) distribution using $\mu = \sqrt{D}, \lambda = D/\gamma$, where $D = \theta(3\gamma + \theta)$ which has applications in insurance, economics, robust statistics and model-based clustering can be found in Punzo (2019) [177], and an extended inverse Gaussian (EIG) distribution in Lai, Ji, and Yan (2020) [125].

10.4   APPLICATIONS

The IGD has applications in modeling and analyzing asymmetric data. For example, it is used to model particle sizes in atmospheric sciences, fluid dynamics, metallurgy, ocean engineering, etc. It can also arise in various contexts like accumulated damage, failure growth, service time of employees, hospital stays of patients of a specific disease, re-order times of some inventory items, etc., all of which involve time as the variable. Other fields of applications include tracer distribution in biological systems, reliability (wear-outs of components and parts in cumulative damage processes), Internet communications, and quality of drinking water (fluoride and dissolved oxygen levels). If a Weiner process has drift d and diffusion constant σ^2, the first passage time to failure is IGD$(c/d, (c/\sigma)^2)$ where $c > 0$ is the failure cutoff or critical limit. In general, any diffusion process with boundary conditions can be modeled by the IGD (Edgeman and Shanmugam (1990) [61]). Historically, it is more popular in sequential analysis (Wald (1947) [225]).

10.4.1   RELIABILITY

The lifetime of machines, appliance, or components (within appliances and machines) can be modeled using IGD. It is used in reliability engineering to model failure times of components and structures, renewal theory, etc. It is the preferred choice for accelerated life tests, as the hazard function increases and then decreases. Multiple surface faults of bearings rotating in a uniform direction can be approximated as a point process in which inter-event times are approximately IG distributed (Boŝkoski and Juričič (2014) [24]). It also finds applications in survival analysis, fiducial inference, etc. (Jayalath and Chhikara (2020) [101]).

10.4.2   ANALYSIS OF RECIPROCALS

Analysis Of Variance (ANOVA) is a statistical technique to test equality of several means ($H_0 : \mu_1 = \mu_2 \cdots = \mu_k$) by decomposing the total variation of all samples combined into constituent parts (within groups, between groups, etc.). A fundamental assumption in ANOVA is that the samples come from normal populations. This assumption may not always hold. An alternative in such situations is the Analysis Of Reciprocals (ANOR) or Reciprocal ANOVA (R-ANOVA) in which a relation between main effects and interactions is established using reciprocals of data values under the assumption of a linear model for reciprocal mean[3] (Tweedie(1957) [219]). More specifically, let (y_i, x_i) for $i = 1, 2, \ldots, n$ denote the failure time and corresponding stress level, where $y_i \sim$ IG(μ_i, λ) are independent. It is assumed in the simplest ANOR one-way classification that the n_i items in the ith sample are drawn from IG(μ_i, λ) for $i = 1, 2, \ldots, m$. From the identity

$$\sum_{i=1}^{m} \sum_{j=1}^{n_i} (1/x_{ij} - 1/\overline{x}_{..}) = \sum_{i=1}^{m} (1/\overline{x}_{i.} - 1/\overline{x}_{..}) + \sum_{j=1}^{n_i} (1/\overline{x}_{.j} - 1/\overline{x}_{..}), \qquad (10.28)$$

it follows that the total sum of residuals of reciprocals can be partitioned into two components in the single factorial design setup. Tweedie showed that these terms follow scaled-χ^2 distribution with respective DoF $\sum_i n_i - 1$, $m - 1$, and $\sum_i n_i - m$ with scaling factor $(1/\lambda)$, and the two RHS terms are independent so that the F-test can be used as in the case of ANOVA. Similarly, the homogeneity of λ (which is analogous to the homoscedasticity of variance in classical linear models) can also be tested using ANOR procedures (Seshadri(1999)) [198]. This model can be extended to higher-factorial experiments by assuming that the samples are drawn from IG$(\mu + \alpha_i + \beta_j, \lambda)$ for $i = 1, 2, \ldots, m$, $j = 1, 2, \ldots, n$ such that $\mu > 0$, $\sum_i \alpha_i = \sum_j \beta_j = 0$.

10.5   SUMMARY

This chapter introduced the inverse Gaussian distribution, and its basic properties. These distributions are extensively used in engineering and scientific fields in lieu of gamma and Weibull distributions. Several extensions of these distributions are available in the literature. However, we have described only the most important ones due to space limitations.

---

[3]This is called reciprocal linear regression model $\mu^{-1} = \alpha + \beta x + \epsilon$.

CHAPTER 11

Birnbaum–Saunders Distribution

11.1 INTRODUCTION

The Birnbaum–Saunders distribution (BSD) was first introduced by Fletcher (1911) [69], and in a slightly different parametric form by Freudenthal and Shinozuka (1961) [72]. It became popular with Birnbaum and Saunders (1969) [21], whose work originated from investigations involving crack-size growth in materials made of metals or alloys caused by cyclic vibrations. Duration of the cycle can range from a few seconds to hours, days, or even months depending on the field of application. For example, cracks in buildings and bridges are caused by seismic events (quakes and aftershocks), which could be rare in some locations. Cracks may appear casually, and advance to larger sizes randomly by a non-negative amount (called crack extensions), until it culminates in a *critical size*. The affected part may pose a threat to safe operation after a discrete number of cycles N, upon which a fatigue failure occurs either by a dominant crack reaching a critical size or a set of spatially correlated cracks reaching a threshold level. A crack is considered to be dominant either due to its size, orientation, or its sensitive location or a combination of them. Hence, even small cracks in critical or sensitive regions are dominant. Examples are cracks or ruptures on tankers that transport highly inflammable products or toxic materials, cracks on internal combustion engines, boilers and storage tanks used in industry, and vacuum-operated machinery, because even a small crack can have disastrous consequences. Similarly, cracks on medical devices of various sorts could pose health hazards to operators and patients. It is implicitly assumed that the crack growth during each cycle $(1 \dots N)$ is independent, and have the same statistical distribution. It is known as Miner's Rule in materials physics, where the critical damage caused by stress levels after n cycles is proportional to n/N.

There are in general two models called *fatigue failure* and *stochastic wear-out failure* (of which the former is a particular case) depending on the device, part, component, or structure. The fatigue failure (due to stress, tensions, etc.) under cyclic loadings in manufacturing and transportation systems can be modeled by BSD,[1] from which it gets the name *fatigue-life distribution*. BSD along with lognormal, Weibull, and gamma distributions are used in life-time modeling of components

---

[1]This is called the force of mortality in some fields where forces are physical or mechanical.

with specified wear. The PDF of two-parameter BSD is

$$f(x; \alpha, \beta) = \frac{1}{2\sqrt{2\pi}\,\alpha\beta}(\beta/x)^{1/2}(1 + \beta/x)\,\exp(1/\alpha^2)\,\exp\{-\frac{1}{2\alpha^2}\,(x/\beta + \beta/x)\}, \quad (11.1)$$

for $x, \alpha, \beta > 0$. The CDF or SF can be expressed in terms of standard normal CDF as

$$F(x; \alpha, \beta) = \Phi\left([\sqrt{x/\beta} - \sqrt{\beta/x}]/\alpha\right), \quad S(x; \alpha, \beta) = \Phi\left([\sqrt{\beta/x} - \sqrt{x/\beta}]/\alpha\right). \quad (11.2)$$

Differentiate (11.2) wrt x to get the PDF as

$$f(x; \alpha, \beta) = \phi\left([\sqrt{x/\beta} - \sqrt{\beta/x}]/\alpha\right) \times (1/\alpha)\left(1/(2\sqrt{\beta}\sqrt{x}) + \sqrt{\beta}/(2x^{3/2})\right). \quad (11.3)$$

Take $2\sqrt{x}$ as a common factor from the denominator of (11.3) to get

$$f(x; \alpha, \beta) = \phi\left([\sqrt{x/\beta} - \sqrt{\beta/x}]/\alpha\right)\left[1/\sqrt{\beta} + \sqrt{\beta}/x\right]/(2\alpha\sqrt{x}). \quad (11.4)$$

Rearrange to get another form as

$$f(x; \alpha, \beta) = \phi\left([\sqrt{x/\beta} - \sqrt{\beta/x}]/\alpha\right)\left[(\beta/x)^{1/2} + (\beta/x)^{3/2}\right]/(2\alpha\beta). \quad (11.5)$$

It is denoted by BS(α, β) where α is the shape and β is the scale parameter.

11.1.1   DERIVATION OF BSD

Consider a nearly smooth material (made of metals, alloys, fiber, plastics, or concrete) that undergoes cycles of stress loads continually or periodically (as in vibrations during landing and takeoff of aircrafts that continuously operate with a break in-between) that results in a cumulative degradation process on the overall quality. The shape of the material is unimportant to the formulation, but could be flat or curved surfaces (as in fuselage of aircrafts and submarines, train-tracks, conveyor belts, tyres, surface of buildings and bridges, blades of turbines; circular as in brake pads, discs of various sorts; spherical as in ball-bearings; cylindrical as in shafts of rotating machinery, elevator cables; or a combination of them as in piston pumps, engines, gearboxes, complex joints, crankshafts, etc.). A clearly demarcated region must also be identified for each crack as in buildings, railway tracks, boilers, and turbines, where it could appear at multiple locations.

Birnbaum and Saunders (1969) considered the distribution of a nonlinear transformation of Gaussian variate as

$$X = \beta[\alpha Z/2 + \sqrt{(\alpha Z/2)^2 + 1}]^2, \quad (11.6)$$

where $Z \sim N(0,1)$, α and β are positive constants (in comparison, the IGD is related to the standard normal as $Y = \sqrt{\lambda/X}(X/\mu - 1)$).

The genesis of the fissure or crack differs among various applications, but must be well-understood because its growth process could differ depending on the context (type of stress like excess vibrations, mechanical overloads, axial and bending loads, extra rotational speeds or pressure, tensions and torsions, unstable voltage induced stress, etc.) as well as material orientations (as in off-shore vessels, rigs and platforms subjected to cyclic wave loadings, rotating parts of a machinery like windmills and turbine fans). It is assumed that fatigue failure occurs as a result of repeated application of cyclic stress patterns, due to which a dominant crack grows incrementally until it culminates in a critical size beyond which it becomes useless or unsafe without repair or replacement. Meanwhile, accelerated life testing is carried out when the material of interest is durable and reliable, so as to obtain failure data in as little time as possible.[2] If the increase in crack-size X_j at the j-th cycle is assumed to be a random variable with mean μ_j and variance σ_j^2, the overall size $X = \sum_j X_j$ is approximately normally distributed using central limit theorem. Then the probability that the crack-size does not exceed a critical threshold w in minimum number of cycles is given by

$$\Pr[X \le w] = \Phi((w - n\mu_0)/(\sqrt{n}\sigma_0)) = \Phi((w/\sqrt{n}\sigma_0 - \mu_0\sqrt{n}/\sigma_0)). \qquad (11.7)$$

It is assumed that $\Pr[X < 0]$ is negligible. Equation (11.7) could also be written as $\Phi((w - n\mu_0)/(\sqrt{n}\sigma_0)) = \Phi(\sqrt{w\mu_0}/\sigma_0 \left(\sqrt{(w/\mu_0)/n} - \sqrt{n/(w/\mu_0)} \right))$. The CDF of T (lifetime until failure) is $1 - \Pr[X \le w] = \Phi(\mu_0\sqrt{t}/\sigma_0 - w/(\sqrt{t}\sigma_0))$. Write $U = \mu_0\sqrt{t}/\sigma_0 - w/(\sqrt{t}\sigma_0)$. Square it and solve for T to get $T = (w/\mu_0)[U\sigma_0/(2\sqrt{\mu_0 w}) + ((U\sigma_0/(2\sqrt{\mu_0 w}))^2 + 1))^{1/2}]^2$. This is similar to (11.6) with $\beta = w/\mu_0$ and $\alpha = \sigma_0/(\sqrt{\mu_0 w})$.

Cracks may develop at two (or more) diagonally opposite edges of an object as in the case of railway tracks, windshields of vehicles, screens, flat belts, pipes and cylindrical shafts, etc. A critical condition will soon be reached when the cracks advance towards each other direction (along the surface). If X and Y denote two such crack evolutions, the lifetime of the system can be modeled as a convolution of X and Y.

Problem 11.1 If $Z \sim N(0,1)$, find the distribution of $X = \beta[\alpha Z/2 + \sqrt{\alpha|Z|/2 + 1}]^2$, and $Y = \beta[\alpha Z/2 + (\alpha|Z|/2 + 1)^{1/p}]^p$ for $p > 1$.

11.1.2 ALTERNATE REPRESENTATIONS

Write $(1 + \beta/x) = (x + \beta)/x$, and expand $\phi(x)$ to get the form

$$f(x; \alpha, \beta) = [1/(2\sqrt{2\pi})] [(x + \beta)/(\beta\sqrt{\alpha x^3})] \exp(1/\alpha^2) \exp\{-\frac{1}{2\alpha^2} (x/\beta + \beta/x)\}. \qquad (11.8)$$

This can also be written as

$$f(x; \alpha, \beta) = 1/(2\alpha\beta)(x/\beta)^{-.5}(1 + (x/\beta)^{-1}) \exp(1/\alpha^2)/\sqrt{2\pi} \exp\{-\frac{1}{2\alpha^2} (x/\beta + \beta/x)\}, \qquad (11.9)$$

---

[2]The accelerated life models relate the stress or loads to the hazard rate, where the stress can be constant, cyclic, random, step-wise increasing, or progressive.

for $x > 0, \alpha, \beta > 0$, or as

$$f(x; \alpha, \beta) = (1/(2\sqrt{\pi}\alpha^2\beta^2))\,((x^2 - \beta^2)/\sqrt{x/\beta} - \sqrt{\beta/x})\,\exp(2/\alpha^2)$$

$$\exp\{-\frac{1}{2\alpha^2}\,(x/\beta + \beta/x)\}. \tag{11.10}$$

Write $t(x) = x^{1/2} - x^{-1/2}$. Then the PDF becomes

$$f(x; \alpha, \beta) = 1/(2\sqrt{2\pi}\alpha\beta)t(x/\beta)\exp(-[t(x/\beta)]^2/(2\alpha^2)). \tag{11.11}$$

It has mean $\mu = \beta(1 + \alpha^2/2)$ and variance $(\beta\alpha)^2\,(1 + 5\alpha^2/4)$. As the median is β, it is applicable in those situations where the sample median is a better estimate than the sample mean. The difference between mean and median is linear in β and quadratic in α, so that the skewness rapidly increases for fixed β when α values are greater than one, and decreases otherwise. Thus, it approaches near symmetry around the median β as $\alpha \to 0$. These properties make BSD an excellent choice, in place of Gaussian law, in statistical models where error terms deviate from normality. As β is a scale parameter, it follows that $X/\beta \sim BS(\alpha, 1)$. Desmond (1986) [57] represented the PDF as a mixture of an IG distribution and a Reciprocal IG (RIG) distribution as $f(x; \alpha, \beta) =$

$$\frac{1}{3}\left[(\frac{\beta}{2\pi\alpha^2x^3})^{1/2}\exp\left(\frac{1}{2\alpha^2\beta(x - 2\beta + \beta^2/x)}\right) + (\frac{1}{2\pi\alpha^2\beta x})^{1/2}\right.$$

$$\left.\exp\left(-\frac{\beta}{2\alpha^2(\frac{1}{x} - \frac{2}{\beta} + \frac{x}{\beta^2})}\right)\right]. \tag{11.12}$$

A three-parameter BSD has PDF $f(x; \alpha, \beta, \mu) =$

$$\frac{[(x - \mu)/\beta]^{1/2} + (\beta/(x - \mu))^{1/2}}{2\alpha(x - \mu)}\phi\left([((x - \mu)/\beta)^{1/2} - (\beta/(x - \mu))^{1/2}]/\alpha\right), \tag{11.13}$$

for $x > 0, \alpha, \beta > 0$, where β is the scale, α is the shape, and μ is the location parameter. This reduces to the standard BSD when $\mu = 0$ and $\alpha = \beta = 1$ as

$$f(x; 1, 1) = (\sqrt{x} + \sqrt{1/x})/(2\sqrt{2\pi}x)\phi(\sqrt{x} - \sqrt{1/x}). \tag{11.14}$$

The transformation $Y = \beta/X$ results in the PDF

$$f(y; \alpha, \beta) = y^{-3/2}(1 + y)\exp((1 + y^2)/2\alpha^2)\exp(1/\alpha^2)/(2\alpha\sqrt{2\pi}). \tag{11.15}$$

Problem 11.2   If $X \sim Z(0, 1)$ is a standard normal random variable, find the distribution of $Y = (b/2)[a^2X^2 + aX\sqrt{a^2X^2 + 4} + 2]$.

Problem 11.3   Prove that the mean (μ) of $BS(\alpha, \beta)$ is always greater than the median, and find $\Pr[median \leq X \leq \mu]$.

11.2   RELATION TO OTHER DISTRIBUTIONS

If $Z \sim N(0,1)$ is a unit normal variate, then $X = \beta[\alpha Z/2 + \sqrt{(\alpha Z/2)^2 + 1}]^2$ has a BS distribution with shape parameter α and scale parameter β. Alternately, if $X \sim BS(\alpha, \beta)$, then $Y = \frac{1}{\alpha}\left[(x/\beta)^{1/2} - (x/\beta)^{-1/2}\right] \sim N(0, 1)$. As the square of a standard normal variate has a χ_1^2 distribution, it follows that $T = Y^2 = ((x/\beta) + (\beta/x) - 2)/\alpha^2 \sim \chi_1^2$. As it is a nonlinear transformation of the normal variate, some of the Gaussian properties are inherited by the BSD. It was mentioned in Chapter 10 that if $X \sim IGD(\mu, \lambda)$ then $Y = \lambda(X - \mu)^2/(\mu^2 X)$ has (χ_1^2) distribution, so that $T = \sqrt{Y} = \lambda(X - \mu)/(\mu\sqrt{X}) = (1/\sqrt{\mu})\sqrt{X/\mu} - \sqrt{\mu/X}$ has (χ_1) distribution. Thus, it can be obtained as an equal mixture of an IGD and its reciprocal (Desmond (1986) [57]), as also an approximation to the IGD (Bhattacharyya and Fries (1982) [20]). Replacing Z in (11.6) by a truncated-normal, half-normal, and skew-normal variates results in new distributions (Martinez-Florez, et al. (2019) [145]). The distribution of integer part (Chattamvelli and Shanmugam (2021) [47], p. 32 and p. 122) can be used to find a discrete analogue of BSD by expressing the discrete PDF as $f(y) = \Pr[y \le X < y + 1] = S(y + 1) - S(y)$ where $S(x)$ denotes the survival function (SF).

If $X \sim BSD(\alpha, \beta)$ then $Y = \log(X)$ has Sinh-Normal distribution $SinhND(\alpha, \log(\beta), \sigma = 2)$ using the fact that $t(x) = \sqrt{x} - 1/\sqrt{x} = \sinh(\ln(x))$ (Rieck and Nedelman (1991) [184], Jones (2012) [109]). This has CDF $F_y(y) = \Phi((2/\alpha)\sinh((y - \ln(\beta))/2))$. Alternately, if $Z \sim N(0,1)$, then $\mu + \sigma \operatorname{arcsinh}(\alpha Z/2)$ has a sinh-normal distribution, which is symmetric and has kurtosis dependent on α. Thus, the BSD belongs to the log-symmetric family. As $\log(1/X) = -\log(X)$, these distributions have a characteristic property that X and $1/X$ are identically distributed (Jones (2008) [107]). It is unimodal and leptokurtic for $\alpha < 2$, platykurtic for $\alpha = 2$, and bimodal for $\alpha > 2$ (see Figure 11.2). The range of the distribution rapidly increases for increasing α values, so that it can be used to model a variety of practical data analysis problems without data transformations. As it is bimodal for $\alpha > 2$, it can be used to describe processes or events that are bimodal (like COVID-19 first and second wave). See Maehara et al. (2021) [142] for extensions of Sinh distribution.

Problem 11.4   If $X \sim BS(\alpha, \beta)$ prove that the distribution of $Y = 1/X$ is $BS(\alpha, 1/\beta)$.

Problem 11.5   If $X_i \sim BS(\alpha, \beta_i)$ for $i = 1, 2$ are IID, find the distribution of (i) $Y = X_1 + X_2$ and (ii) $Z = 1/X_1 + 1/X_2$.

11.3   PROPERTIES OF BS DISTRIBUTION

It is an asymmetric, unimodal, and positively skewed distribution that approaches symmetry around β as α tends to one (see Figure 11.1). It resembles an inverse exponential distribution (Chattamvelli and Shanmugam (2021) [47], p. 29) for large β values. As the range of the variate is $x > 0$ it fits only positive data. Both the shape and scale parameters are positive as well. The hazard function (HF) defined as $h(x) = f(x)/S(x) = -S'(x)/S(x) = -(\partial/\partial x)\log(S(x))$ of BSD and IG distributions are very similar, but the former HF tends to a constant $1/(2\alpha^2\beta)$ as $x \to \infty$ (Kundu, Kannan, and Balakrishnan (2008) [123]). As mentioned above, the BSD is an equal weight mixture of an IG and an RIG distributions. Analogous to the lognormal distribution, we could define

$$f(x, a, b) = 1/(2 \sqrt{2\pi}ab)(b/x)^{1/2}(1 + b/x) \exp(1/a^2) \exp(-[1/(2a^2)](x/b + b/x))$$

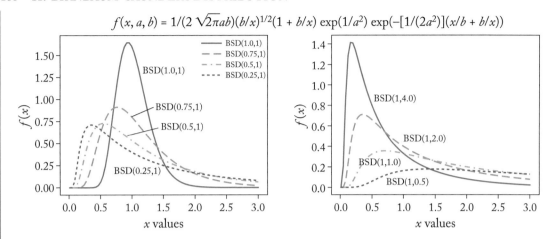

Figure 11.1: Birnbaum–Saunders distributions.

$$f(x, a) = \phi(2 \sinh(x)/a) * 2 \cosh(x)/a$$

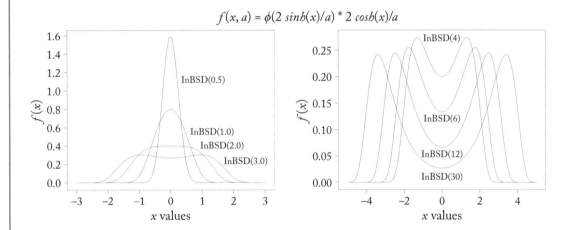

Figure 11.2: Sinh-Normal (log-BSD) distributions.

a log-Birnbaum–Saunders (LBS) distribution, which is a special case of sinh-normal distribution (see Figure 11.2). The CDF of BS(α, β) is related to the error function as

$$F(x; \alpha, \beta) = \Phi\left([\sqrt{x/\beta} - \sqrt{\beta/x}]/\alpha\right) = \frac{1}{2}\left[1 + \text{erf}((\sqrt{x/(2\beta)}) - \sqrt{2\beta/x})/\alpha)\right]. \quad (11.16)$$

This can also be written in terms of confluent hypergeometric function using $\text{erf}(x) = 2x \exp(-x^2)/\sqrt{\pi} \quad {}_1F_1(1, 3/2, x^2)$. As $\Phi(0) = 0.50$, $x = \beta$ is the median. Write $t(x) = \sqrt{x} - 1/\sqrt{x}$, using which the CDF becomes $\Phi(t(x/\alpha)/\alpha)$. Similarly, write $t_1(x) = \sqrt{x} + 1/\sqrt{x}$ and $t_2(x) = \sqrt{x} - 1/\sqrt{x}$, using which the PDF becomes $t_1((x - \alpha)/\beta)\phi(t_2((x - \alpha)/\beta)/\alpha)$.

It has a scaling property similar to that of IG and RIG distributions. If $X \sim \text{BSD}(\alpha, \beta)$ then $cX \sim \text{BSD}(\alpha, c\beta)$ where $c > 0$. Thus, it is closed under scalar multiplication when the multiplier is a positive real number. The reciprocal $Y = 1/X$ of $\text{BSD}(\alpha, \beta)$ is distributed as $\text{BSD}(\alpha, 1/\beta)$, using which we could easily find the inverse moments (see below). Thus, it is also closed under reciprocation.

Problem 11.6   If $X \sim \text{BS}(\alpha, \beta)$ find the distribution of $Y = 1/(cX)$ where $c > 0$ is a positive real number.

The relationship between standard normal and BSD allows us to express the quantiles of BSD in terms of those of standard normal as $X_p = (\beta/4)\left(\alpha Z_p + \sqrt{4 + \alpha^2 z_p^2}\right)^2$, for $0 < p < 1$. This representation is also helpful to extend BSD to higher dimensions. For example, the bivariate BSD can be defined using its CDF as

$$F(x, y; \alpha_1, \alpha_2, \beta_1, \beta_2, \rho) = \Phi_2\left([\sqrt{x/\beta_1} - \sqrt{\beta_1/x}]/\alpha_1, [\sqrt{y/\beta_2} - \sqrt{\beta_2/y}]/\alpha_2, \rho\right), \quad (11.17)$$

where $\Phi_2()$ is the CDF of bivariate normal distribution with correlation ρ.

As the mean $\geq$ median $\geq$ mode for right-skewed distributions, the mode is less than β. However, explicit expressions for mode exist only for special values of the parameters. For example, the mode is the root above 1 of the cubic equation $x^3 + (\alpha^2 + 1)x^2 + (3\alpha^2 - 1)x - 1 = 0$ when $\beta = 1$. The MD can be found using the theorem given in Chapter 1 of Part I (ref. [47]) as

$$\text{MD} = 2 \int_0^{\beta(1+\alpha^2/2)} \Phi\left((1/\alpha)[\sqrt{x/\beta} - \sqrt{\beta/x}]\right) dx. \quad (11.18)$$

11.3.1   MOMENTS AND GENERATING FUNCTIONS

The kth moment is most easily found from $E(x/\beta)^k = E\left(\alpha Z/2 + \sqrt{(\alpha Z/2)^2 + 1}\right)^{2k}$ using binomial expansion as

$$E(x^k) = \beta^k \sum_{j=0}^{k} \binom{2k}{2j} \sum_{i=0}^{j} \binom{j}{i} \frac{2(k + i - j)!}{2^{k+i-j}(k + i - j)!}(\alpha/2)^{2(k+i-j)}, \quad (11.19)$$

whereas the kth fractional moment is $\alpha'_k = \beta^k I(k, \alpha)$ where $I()$ is the Bessel function (Rieck (1999) [182]). The first few moments are $E(X) = \mu = (\alpha^2 + 2)\beta/2$, $\sigma^2 = (5\alpha^2 + 4)\alpha^2\beta^2/4$ (Table 11.1). Ordinary moments of BSD (and truncated BSD) can be expressed in terms of moments of $Z(0,1)$ using the above technique.

Problem 11.7   Obtain an expression for the k-th moment of BSD in terms of corresponding moments of standard normal random variate by expanding $X^k = \beta^k[\alpha Z/2 + \sqrt{(\alpha Z/2)^2 + 1}]^{2k}$ using binomial theorem.

Table   11.1:   Properties   of   BSD $(f(x;\alpha,\beta) = 1/(2\sqrt{2\pi}\alpha\beta)(\beta/x)^{1/2}(1 + \beta/x)\exp(1/\alpha^2)\exp\{-\frac{1}{2\alpha^2}(x/\beta + \beta/x)\})$

Property	Expression	Comments
Range of X	$x \geq 0$	Continuous
Mean	$\mu = \beta(1 + \alpha^2/2)$	$\to \beta$ as $\alpha \to 0$
Median	β	
Variance	$\sigma^2 = (\alpha\beta)^2(1 + 5\beta^2/4)$	
μ_3'	$\beta^3(1 + 9\alpha^2/2 + 9\alpha^4 + 15\alpha^6/2)$	$E[X^3]$
Skewness	$\gamma_1 = 4\alpha(11\alpha^2 + 6)/(5\alpha^2 + 4)^{3/2}$	+ve skewed
Kurtosis	$\beta_2 = 3 + 6\alpha^2(93\alpha^2 + 40)/(5\alpha^2 + 4)^2$	Leptokurtic
CV	$(\alpha/(\alpha^2 + 2))(5\alpha^2 + 4)^{1/2}$	Indep. of β
CDF	$\Phi\left((1/\alpha)\left[\sqrt{\frac{x}{\beta}} - \sqrt{\frac{\beta}{x}}\right]\right)$	$0 < x < \infty$
Inv. CDF	$(\beta/4)[\alpha\Phi^{-1}(p) + \sqrt{4 + (\alpha\Phi^{-1}(p))^2}]^2$	quantiles
MGF	$(1/2)\exp(\delta - (\delta^2 - 2\beta t)^{1/2})[1 + 1/(1 - 2\alpha t/\delta)^{1/2}]$	$\delta = \beta/\alpha$
CGF	$\delta(1 - [1 + 2\alpha^2 it/\beta]^{1/2})$	
ChF	$(1/2)\exp(\delta - (\delta^2 - 2\beta it)^{1/2})[1 + 1/(1 - 2\alpha it/\delta)^{1/2}]$	$\delta = \beta/\alpha$
Additivity	$X_i \sim BS(\alpha,\beta_i) \Rightarrow \sum_i X_i \sim BS(\alpha, \sum \beta_i)$	Indep.

Approaches normality as $N(\beta, \alpha^2\beta^2)$ for very small α values.

As the coefficient of skewness is $\gamma = 4\alpha(11\alpha^2 + 6)/(5\alpha^2 + 4)^{3/2}$, which is independent of the scale parameter β. This approaches zero as $\alpha \to 0$. As the mean $\mu = \beta(1 + \alpha^2/2) \to \beta$ as $\alpha \to 0$, the distribution approaches Dirac's δ function centered at β.

Problem 11.8   If $X \sim BS(\alpha,\beta)$, prove that $E(X/\beta) = 1 + \alpha^2/2$ and $E((X/\beta)^2) = 3\alpha^4/2 + 2\alpha^2 + 1$.

Writing $(5\alpha^2 + 4)^{3/2} = 11.180339887\alpha^3(1 + 4/(5\alpha^2)^{3/2})$ shows that the skewness approaches 3.9354796 as $\alpha \to \infty$. The kurtosis coefficient is $\gamma_2 = 3 + 6\alpha^2(93\alpha^2 + 40)/(5\alpha^2 + 4)^2$, which also is independent of β. This approaches 3 as $\alpha \to 0$ and 3+22.32 = 25.32 as $\alpha \to \infty$. A normal approximation is appropriate to BSD as $N(\beta, \alpha^2\beta^2)$, and to RBSD as $N(1/\beta, \alpha^2/\beta^2)$ for very small α values. As the reciprocal of a BSD has an identical distribution, it is easy to find the inverse moments from the relationship $X \sim BS(\alpha,\beta) \Rightarrow Y = 1/X \sim BS(\alpha, 1/\beta)$ by replacing β by $1/\beta$ in (11.19). Alternatively, as $Y = X/\beta \sim BS(\alpha, 1)$, it follows that $1/Y \sim BS(\alpha, 1)$ so that

$E(X^{-k}) = E(X^k/\beta^{2k})$. Thus, all inverse moments exist as their positive counterparts exist. In particular, the first inverse moment is $(1 + \alpha^2/2)/\beta$.

Problem 11.9   If $X \sim BS(\alpha, \beta)$, prove that $E(X/\beta + \beta/X) = 2 + \alpha^2$ and Var $(X/\beta + \beta/X) = 2\alpha^4$.

The ChF is

$$\phi_x(t) = \int_0^\infty \exp(itx)\phi\left([\sqrt{x/\beta} - \sqrt{\beta/x}]/\alpha\right)\left[(\beta/x)^{1/2} + (\beta/x)^{3/2}\right]/(2\alpha\beta)dx. \quad (11.20)$$

Transform the variable using $Y = X/\beta$ and split the integration into two parts to get

$$\phi_x(t) = \frac{1}{2}\left([1 + 1/\sqrt{1 - \delta i}] \quad \exp([1 - \sqrt{1 - \delta i}]/\alpha^2)\right), \quad \text{where} \quad \delta = 2\alpha^2\beta. \quad (11.21)$$

(Leiva (2016) [128]). The MGF is given by

$$(1/2)\exp(\delta - (\delta^2 - 2\beta t)^{1/2})[1 + 1/(1 - 2\alpha t/\delta)^{1/2}], \quad \text{where} \quad \delta = \beta/\alpha. \quad (11.22)$$

As $x > 0$, the random variable $Y = \ln(X)$ has ChF $\phi_y(t) = E(\exp(ity)) = E(\exp(it \ln(x))) = E(\exp(\ln(x^{it}))) = E(x^{it})$. As t is a real dummy variable, all fractional moments also exist.

11.4   TAIL AREAS

As noted above, the CDF is expressible in terms of the standard normal CDF as

$$F(x; \alpha, \beta) = \Phi\left(\left[\sqrt{\frac{x}{\beta}} - \sqrt{\frac{\beta}{x}}\right]/\alpha\right), \quad (11.23)$$

where $\Phi()$ is the CDF of Z(0,1). The CDF can also be expressed in terms of error function as shown above. Approximations to the right-tail areas can be developed as the distribution tails off slowly to the right (Figure 11.1).

11.5   FITTING

Moment estimates of the parameters are obtained by equating the sample moments with corresponding population counterparts. Let m be the sample mean and s^2 the sample variance for the two-parameter BSD. Then $m = \mu = \beta(1 + \alpha^2/2)$, and $S = s^2 = (\beta\alpha)^2 (1 + 5\alpha^2/4)$. Square the first expression and divide by the second to get the quartic equation $\alpha^4(5m^2 - S) + 4\alpha^2(m^2 - 4) - 4S = 0$. From this we get $\hat{\alpha}^2/2 = [(4 - m^2) + \sqrt{(4 - m^2)^2 + S(5m^2 - S)}]/(5m^2 - S)$ and $\hat{\beta} = m/(1 + \hat{\alpha}^2/2)$, which exists only when the CV is greater than $\sqrt{5}$. Instead of equating the means and variances, one could equate the median and variance to get an immediate estimate of β as $\hat{\beta} = \text{median } (x_1, x_2, \ldots, x_n)$, and $5\hat{\alpha}^2 = 2[(1 + 5S/\hat{\beta})^{1/2} - 1]$ where S is the sample variance (From and Li (2006) [73]).

The modified method of moments uses the fact that if $X \sim \mathrm{BS}(\alpha, \beta)$ then $Y = 1/X \sim \mathrm{BS}(\alpha, 1/\beta)$. Thus, equating the first moment and inverse moment gives $m = \mu = \beta(1 + \alpha^2/2)$, and $1/\hat{x} = (1 + \alpha^2/2)/\beta$, where $\hat{x}$ is the HM. Solve simultaneously to get $\hat{\alpha}^2 = 2(\sqrt{(m/\hat{x})} - 1)$ and $\hat{\beta} = \sqrt{m\hat{x}}$, which always exist (Ng, Kundu, and Balakrishnan (2003) [166]). As $E[(X/\beta)^{1/2} - (\beta/X)^{1/2}] = 0$, we get the identity $\sum_{j=1}^{n}(x_j/\beta)^{1/2} = \sum_{j=1}^{n}(\beta/x_j)^{1/2}$. Take constants outside the summation and simplify to get $\hat{\beta} = \sum_{j=1}^{n} x_j^{1/2} / \sum_{j=1}^{n}(1/x_j)^{1/2}$ (Johnson, Kotz, and Balakrishnan (2005) [104]).

The MLE estimates are found from log-likelihood function $\ln(L(x_1, \ldots, x_n, \alpha, \beta)) =$

$$-n[\ln(\alpha) + \ln(\beta)] + \sum_{j=1}^{n} \ln\left[(\beta/x_j)^{1/2} + (\beta/x_j)^{3/2}\right] - (1/(2\alpha^2)) \sum_{j=1}^{n}(x_j/\beta + \beta/x_j - 2),$$

(11.24)

where $x_1, x_2, \ldots, x_n$ is a random sample of size $n > 1$ from the BSD in which none of the values are zeros (zero values and all data values too close to zero must be discarded to avoid memory overflow problems). Such small values in the sample are more likely for increasing α values.

Differentiate wrt α, and equate to zero to get $\alpha^2 = (\overline{x}/\beta + \beta/\hat{x} - 2)$ where $\hat{x} = 1/\sum_{j=1}^{n}(1/x_j)/n$ is the harmonic mean (HM) of the sample, which makes sense as none of the sample values are negative, and zero values has been discarded. Differentiate wrt β, substitute for α^2, and equate to zero to get $\beta^2 - 2\beta(2\hat{x} + \psi(\beta) + \hat{x}(\overline{x} + \psi(\beta))) = 0$, where $\psi(\beta) = 1/[\sum_{j=1}^{n}(1/(\beta + x_j))]$, which is $(1/n)$ times the HM of the displaced data values. Note that this is not a quadratic equation in β because $\psi(\beta)$ appears twice in the equation. Hence, a nonlinear equation solver must be used to find the positive root. Birnbaum and Saunders (1969) [22] proved that the unique positive root satisfies $\hat{x} < \hat{\beta} < \overline{x}$. As the AM $\geq$ GM $\geq$ HM, the GM of the sample values can be used as an initial guess, instead of starting with an arbitrary initial value. Alternately, the "mean-mean" which is the GM of AM and HM defined as $[(\sum_{j=1}^{n} x_j)/(\sum_{k=1}^{n}(1/x_k))]^{1/2}$ can be used (Johnson, Kotz, and Balakrishnan (2005) [104]).

The MLE of α^2 is found from the MLE of β using $\hat{\alpha}^2 = (\overline{x}/\hat{\beta} + \hat{\beta}/\hat{x} - 2)$. As α is positive, take the positive square root to get the MLE of α. These are consistent estimators for the unknowns as shown by Birnbaum and Saunders (1969) [22]. Balakrishnan and Zhu (2014) [16] used expected value of ratios of IID BSD random variables to come up with estimators $\hat{\alpha}^2 = 2(\sqrt{\overline{z}} - 1)$ and $\hat{\beta} = m/\sqrt{\overline{z}}$ where $\overline{z} = \frac{1}{n(n-1)} \sum_{i \neq j=1}^{n}(x_i/x_j)$ where none of the sample values are zero or too close to zero. If the sample values are first sorted in ascending order, the above sum can be computed by summing the below-diagonal or above-diagonal entries of the matrix formed by the ratio of sorted sample values, so that $\overline{z} = \frac{2}{n(n-1)} \sum_{i<j=1}^{n}(x_i/x_j)$. Balakrishnan and Kundu (2019) [14] proposed new estimator using From and Li (2006)[73] estimator $\hat{\beta} = \mathrm{median}(x_1, x_2, \ldots, x_n)$, as $\hat{\alpha}^2 = (1/n) \sum_{j=1}^{n} u_j^2$ where $u_j = (x_j/\hat{\beta})^{1/2} - (\hat{\beta}/x_j)^{1/2}$. They also discuss interval and Bayesian estimators. Inference for the common mean when several samples from possibly different BS distributions with the same mean with an application to the fatigue life of 6061-T6 aluminum coupons is considered by Guo et al. (2017) [84].

11.6   EXTENSIONS OF BS DISTRIBUTION

A size-biased BS (SBBS) distribution can be obtained by finding the expected value of $(a + bx)$ (or equivalently $1 + cx$)) where a, b are non-zero constants. As $E(x) = \beta(1 + \alpha^2/2)$, we get the PDF as $(a + bx) f(x; \alpha, \beta)/(a + b\beta(1 + \alpha^2/2))$, where $f(x; \alpha, \beta)$ is any of the representations given in Section 11.1. Put $a = 0, b = 1$ to get length-biased BSD (LB-BSD). The mean of LB-BSD is $\mu = \beta((2 + 4\alpha^2 + 3\alpha^4)/(2 + \alpha^2)) = \beta(1 + 3\alpha^2(1 + \alpha^2)/(2 + \alpha^2))$ and the mode is the positive root of the cubic $t^3 - \beta(\alpha^2 - 1)t^2 + \beta^2(\alpha^2 - 1)t - \beta^3 = 0$ for $\alpha < 2$, and is bimodal for $\alpha > 2$ (and one repeated root for $\alpha = 2$) (Oliveira et al. (2020) [167]). As the range of the BSD is $x > 0$, a logarithmic BSD can be defined as $Y = \log(X) \sim BS(\alpha, \beta)$ for $x \geq 1$ (Section 11.6.1, page 174). Other simple extensions include transformed, truncated, transmuted, and exponentiated BSD. Replacing Y in the "unit-normal-BSD" relation $Y = \frac{1}{\alpha}\left[(x/\beta)^{1/2} - (x/\beta)^{-1/2}\right]$ by any symmetric continuous distribution (as also symmetrically truncated versions of them including truncated Gaussian) results in new families of distributions (see Leiva et al. (2012) [129] for Student's T-BS distribution). See Teimouri, Hosseini, and Nadarajah (2013) [213] for the ratio of two BSD random variables and its applications.

A three-parameter BSD is obtained by the transformation $Z = (1/a)\left((x/b)^c - (b/x)^c\right)$ where $c > 0$, a is the shape-parameter, and b is the scale-parameter (Diaz–Garcia and Dominguez–Molina (2006) [59]). Another three-parameter BSD was reported by Owen (2006) [168] by relaxing the assumption of independent crack growth and considering critical crack growth as a long-memory process. Cordeiro and Lemonte (2010) [51] proposed the β-Birnbaum–Saunders distribution and illustrated its use by means of three real data sets. Marchant et al. (2013) [143] used it for air contamination modeling, Wanke and Leiva (2015) [226] used it to model demand distribution during lead-time of inventory systems. This reduces to the BSD for $c = 1/2$. The epsilon-generalized BSD is the distribution of $W = (\beta/4)[\alpha X + \sqrt{\alpha^2 X^2 + 4}]^2$ where X has an epsilon skew-symmetric distribution (Arellano–Valle, Gomez, and Quintana (2005) [6]) defined as

$$g(x; \epsilon) = \begin{cases} f(x/(1 + \epsilon)) & \text{for } x < 0; \\ f(x/(1 - \epsilon)) & \text{otherwise.} \end{cases}$$

This has PDF

$$f(x; \alpha, \beta, \epsilon) = (\beta + x)/(2\alpha\sqrt{\beta}x^{3/2}) \times \begin{cases} f(t(x)/(1 + \epsilon)) & \text{for } x < \beta; \\ f(t(x)/(1 - \epsilon)) & \text{otherwise,} \end{cases}$$

where $t(x) = (1/\alpha)[(x/\beta)^{1/2} - (\beta/x)^{1/2}]$ (Castillo, Gomez, and Bolfarine (2011) [34]). Other extensions are possible by replacing Z in (11.6) by truncated normal or half-normal (which is a special case of truncated normal) random variables. See Eugene, Lee, and Famoye (2002) [63] and Cordeiro and Lemonte (2010) [51] for a beta-generalized BSD (BBSD), Lemonte (2013) [134] for Marshall–Olkin extended BSD, Bourguingon et al. (2017) [27] for transmuted BSD, Saulo et al. (2012) [194] for Kumaraswamy BSD, Pescim et al. (2014) [171] for a 5-parameter (4-shape and 1-scale) Kummer-beta BSD (KBBSD) of which the BBSD is a special case. Santos-Neto et al.

(2014) [190] considered a reparameterized BSD and its application to household food expenses in U.S. See Balakrishnan and Kundu (2019) [13, 14] for a comprehensive review of BSD.

11.6.1   LOG BS DISTRIBUTION

If $g()$ denotes the PDF of a symmetric family, a random variable Y defined as

$$f(y; \alpha, \beta) = g((\log(y) - \log(\alpha))/\beta)/(y\beta), \quad \text{for } y > 0, \tag{11.25}$$

is called a two-parameter log-symmetric distribution. A three-parameter random variable Y defined by the relationship $Z = (2/\alpha)\sinh((Y - \mu)/\sigma)$ where $Z \sim N(0,1)$ and $\alpha > 0$ is called the log-symmetric BSD (LSBSD). The CDF is

$$F(y; \alpha, \mu, \sigma) = \Phi((2/\alpha)\sinh((y - \mu)/\sigma)), \tag{11.26}$$

from which the PDF follows as

$$f(y; \mu, \sigma) = (\sqrt{2}/(\alpha\sigma\sqrt{\pi}))\cosh((y - \mu)/\sigma)\exp(-(2/\alpha^2)\sinh^2((y - \mu)/\sigma)), \tag{11.27}$$

From (11.26), it is clear that this distribution is symmetric around μ.

11.7   RANDOM NUMBERS

The relation between BSD and standard normal or χ_1^2 distribution can be used to generate random numbers. As $Z = (1/\alpha)(\sqrt{X/\beta} - \sqrt{\beta/X})$ is distributed as N(0,1), repeated solution of quadratic equations can be used to generate random numbers. Cross-multiply and square the above expression to get $\alpha^2 z^2 = x/\beta + \beta/x - 2$. Re-arrange to get a quadratic in x as $x^2 - (\alpha^2 z^2 + 2)\beta x + \beta^2 = 0$. Thus, repeated solution of quadratic equations with coefficient of x varying is used to generate random numbers. Note that the discriminant of the above equation is $D = \alpha^2 \beta^2 z^2 (4 + \alpha^2 z^2)$, which being greater than zero results in real and distinct roots (equal roots when $z = 0$). Let x_1 and x_2 be the roots. As the sum of the roots is $x_1 + x_2 = \beta(2 + \alpha^2 z^2)$, one of the roots must always be greater than β (which is the median). Meanwhile, the sum of the roots and product of the roots are both positive because $\beta > 0$. There are many approaches to generate random numbers from it. If the roots are distinct, we could either take the average of them or keep one of them and discard the other using another random number (say from U(0,1)). In the later case, we have to rearrange the roots in ascending order, so that the discarded root is equally likely to fall below or above the median.

1. Generate a random number from Z(0,1) and square it (or generate a random number from χ_1^2).

2. Input the parameters $\alpha > 0$ and $\beta > 0$ of BSD, form the quadratic equation $x^2 - (\alpha^2 z^2 + 2)\beta x + b^2 = 0$ and solve it to get x_1 and x_2.

3. If $x_1 \neq x_2$, return the average $(x_1 + x_2)/2$ (alternately, generate a uniform random number u in (0,1) and discard one of the roots using the magnitude of u and return the other root).

Rieck (2003) [183] did a comparison of two random number generators for BSD, and found that random numbers based on $X = \beta[\alpha Z/2 + \sqrt{(\alpha Z/2)^2 + 1}]^2$ to be more efficient. Leiva et al. (2008) [131] gave algorithms for generalized BSD.

11.8 APPLICATIONS

The BSD has applications in various engineering fields involving cumulative degradation of materials and processes. It is used to model the time to failure, cumulative probability of stress damage, and fatigue failure of components or parts that are subjected to cyclic stress patterns. Birnbaum and Saunders (1969) assumed that the incremental crack size advancement X_j at cycle j results in a cumulative damage model on a dominant crack, so that $Y = \sum_{j=1}^{m} X_j$ (combined crack size in m repeated stress cycles) can be approximated by a Gaussian law $N(m\alpha_0, n\sigma_0^2)$. Assuming independence of X_j's, the random variable Y is Gaussian distributed so that $\Pr[Y \leq t] = \Phi((t - n\alpha_0)/(\sqrt{n}\sigma_0)) = \Phi((t/(\sqrt{n}\sigma_0)) - \alpha_0 \sqrt{n}/\sigma_0))$.

It is a suitable choice in those fields involving physical properties of materials (like materials science, physical chemistry, metallurgy, etc.) that undergo cumulative degradation over deterministic cycles of stress. It is used in the analysis of fatigue data of metals or alloys like aluminum, steel, and structures made of them because fatigue is the primary reason for premature failure of structural, mechanical and hydraulic systems. Electronic engineers use it to model metallic flaws in nano-circuits of semiconductor chips and platters, electronic equipment malfunction mechanisms due to circuit faults caused by voltage spikes and excess heat generated, breaking stress of fibrous composite materials containing carbon fibers (Pescim et al. (2014) [171]). Chemical engineers use it to model faults in transport pipelines used by chemical and allied industries, for failures resulting from progressive chemical reactions, diffusions, and electron-migrations. In structural and manufacturing engineering, a device, component, or portion of interest is placed under repeated cycles of stress until a crack or rupture of large size makes it no longer usable. Mechanical and structural engineers use it to model part deformations, welding, or adhesive failures due to extra load that surpass a limit. This is called stress-strength analysis or load-resistance interference in which the strength (x_1) of a component is measured when subjected to a specific stress (x_2), where either or both of them are assumed to be random. The stress may increase due to mechanical loads, extra flow rate (of fluids, gases, etc.), increase in rotational rates due to voltage fluctuations, temperature or pressure variations, or corrosion and plaque buildup inside tubular regions, etc. Other applications include reliability analysis in metallurgical, mechanical, and structural engineering, right-skewed financial data modeling, and in modeling wind-speed distributions of wind-farms (Mohammadi, Alavi, and McGowan (2017) [155]), among many other fields. Lillo et al. (2018) [137] observes that BSD is a better model than classic extreme value distributions for describing seismic events.

Applications in environmental sciences include air quality modeling, pollutants, and toxic materials modeling in enclosed regions like factories and mines, industrial waste disposition, etc., because leakage could result in personal injury or death, economic loss, and environmental damage. It has also been applied in health sciences to model the accumulation of toxic substances in the respiratory system, critical or chronic illnesses caused by multiple risk factors, etc.

Microbiological assays usually have a minimum threshold below which accurate measurements are either infeasible or meaningless, in which case a left-censored sample using a lower detection limit (LDL) is used to represent all such values. Similarly, upper detection limit (UDL) is used in several medical and epidemiological studies in which all values above a threshold are replaced by UDL. Errors in regression models can be assumed to be BS distributed when the sample is censored. These are called tobit-BS regression (De Sousa (2016) [209]).[3]

An LB-BSD with applications in drinking water quality can be found in Leiva et al. (2009). Fang, Zhu, and Balakrishnan (2016) [64] applied it to independent stochastic comparisons of lifetimes of parallel and series systems, D'Anna, Giorgio, and Riccio (2016) [53] used it to model fatigue life of structural components using standard and accelerated life tests. Applications in agricultural engineering can be found in Garcia-Papani et al. (2017) [77].

11.8.1   BS REGRESSION

The BS regression (BSR) results when the shape and scale parameters of a BSD are assumed to vary wrt the applied stress levels. For simplicity, consider the two-parameter $BSD(\alpha, \beta)$. Assume that α and β are dependent on the stress levels so that $Y_k \sim BS(\alpha(x_k), \beta(x_k))$ where $\alpha(x), \beta(x)$ are unknown relations at stress level x among the data values and parameters that need to be estimated from given data. A simplified BSR results when the scale parameter β is a function of the stress loads and shape parameter is independent of it as $Y_k \sim BS(\alpha, \beta(x_k))$. The relationship between the parameters and covariates can take a variety of forms. Examples include linear and log-linear models ($\log(\beta(x_k)) = \beta_0 + \sum_{j=1}^{m} \beta_j x_{jk} + \epsilon_k$, or $\beta(x_k) = \beta_0 + \beta_1 \log(x_k) + \epsilon_k$, polynomial, power and inverse power models (which can be reduced to log-linear form $y = \beta_0 x^{\beta_1} \Rightarrow \log(y) = \beta_0' + \beta_1 x'$), usually without interactions (terms like $x_j x_k$). The MLE technique can be used to estimate BSR parameters when either or both of the parameters depend on applied stress levels (D'Anna, Giorgio, and Riccio (2016) [53]).

See Maehara et al. (2021) [142] for robust nonlinear BSR models, Leão, Leiva, Saulo, and Tomazella (2017) [132] for a fraility regression model with a medical application, Qu and Xie (2011) [178] for log-BSR with censored data, Vanegas and Paula (2016) [222] for an extension of log-symmetric regression models, De Sousa (2016) [209] for a tobit-BS model with a medical application (measles vaccination in Haiti), Santos-Neto et al. (2016) [191] for reparameterized BSR model, and Oliveira, et al. (2020)[167] for a length-biased BSR model applied to meteorological data.

---

[3]The dependent variable is usually censored either on the left, right, or both ends. Zero-censoring is a special case in which negative values are discarded. Errors in tobit models are usually modeled by any log-symmetric distribution.

11.8.2   ACTUARIAL SCIENCES

It is well known in actuarial sciences that the distribution of claim amounts are highly positively skewed. The BSD with or without truncation is one of the popular choices to model claim sizes (amounts) as it can assume a variety of shapes. Truncated models are more appropriate as zero is meaningless and most insurance companies have an upper cap on cost disbursements every year. Actuarial sciences use the PDF given in (10.9) with parameters α and $\phi = \beta/\alpha$. Sometimes a log transformation is applied to verify whether the resulting data can be approximated by a linear model. See Paula et al. (2014) [169] for an application of BSD to insurance.

11.9   SUMMARY

This chapter introduced the Birnbaum–Saunders distribution and its basic properties. These distributions are extensively used in engineering and scientific fields in lieu of gamma and Weibull distributions. Several extensions of these distributions are available in the literature. However, we have described only the most important ones due to space limitations.

CHAPTER 12

Pareto Distribution

12.1 INTRODUCTION

This distribution is named after the Italian economist Vilfredo Pareto (1848–1923), who studied the income distribution of populace during 1897. Although the variable of interest was "tax money" and "wealth" in his original work, it has been extended to many other types of variables like luminosity of stars and other celestial objects in astronomy, size of various sorts (like firm sizes or headcount in management, size of stored files in computing, size of cities within large countries in sociology, extreme ocean wave heights in ocean engineering, size (area) of aegean islands in geography, species size and abundance in zoology, blackout sizes and restoration times of power grids in power transmission engineering, size or area of a region destroyed by natural calamities like forest fires, oil spills in seas in environmental science, oil-and-gas field-size and reserves distribution in petroleum engineering), frequencies (like frequency of occurrence of family names in a country or in telephone directories, frequency of comet visits in astronomy, frequency of replenishment of perishable items in inventory systems), vibrational amplitudes in mechanical engineering, data faults, or error clusters in communications engineering, position errors in global positioning systems (GPS) and sonar-based rescue and repair missions, durations (like time to complete medical procedures or surgical operations, quarantine periods, duration between major calamities like earthquakes or tsunamis, time to fix bugs in very large and complex software systems, etc.), and costs of commodities (like boats and yachts, air planes, and so on). It is also used for size-frequency modeling studies in aquatic and environmental sciences, epidemiology, microbiology, and semiconductor defects modeling. There are two parameters in the classical version as shown below. The PDF is

$$f(x; k, c) = ck^c x^{-(c+1)} = (c/x)(k/x)^c = (c/k)(k/x)^{(c+1)}, \quad x \geq k, \qquad (12.1)$$

where $k, c > 0$ are constants. Here, k is the scale parameter, and c is the shape parameter (also called tail index, Pareto index, or Pareto exponent[1]). The shape parameter c is called "income inequality parameter" in economics and finance. It should be greater than 1 for finite moments. Its value decides the shape of the tail decay. It is near 2.0 in developed countries for income and wealth distributions, whereas it is near 1.0 for headcount of firms. In economics and finance, k is the cutoff value (like minimum income) and c is a constant that depends on the field.

It reduces to the standard Pareto law for $k = 1$ with PDF

$$f(x; c) = cx^{-(c+1)}, \quad \text{for } x > 1. \qquad (12.2)$$

[1]As it is related to the power law, it is also called power-law exponent.

The parameter k sets the position of the "left edge" of the distribution[2] as data values that can be observed are greater than k. It is convenient to scale the variate as $Y = X/k$ to obtain a one-parameter Pareto distribution with $x > 1$. A shifted Pareto distribution can be used to model the infectivity of COVID-19 and other similar pandemics. It is customary to write the PDF as

$$f(x; x_m, c) = (c/x)(x_m/x)^c = c x_m^c x^{-(c+1)}, \quad x \geq x_m \tag{12.3}$$

in those applications where the minimum cutoff of the variate is known (say x_m). This is called the Pareto distribution of first kind and denoted as $\mathrm{Par}(k, c)$, Pareto-I(k, c) or as Par-I(k, c) to distinguish between the type 1, 2, and 3 distributions defined below. Put $y = x - k$ to get another representation as

$$f(y; k, c) = (c/k)(1 + y/k)^{-(c+1)}, \quad y \geq 0, \tag{12.4}$$

which is the form used in actuarial sciences, hydrology, and some other fields. A change of origin and scale transformation results in the distribution

$$f(x; k, b, c) = c k^c (x - b)^{-(c+1)}, \quad x > b + k. \tag{12.5}$$

It is sometimes called a negative exponent power function distribution. Most of the empirical Pareto models are right-truncated due to obvious reasons.

12.2   RELATION TO OTHER DISTRIBUTIONS

Pareto distribution is a special case of power-law distribution in which the frequency of occurrence of an event varies as a power of a related attribute of that event. Consider the number of files on the hard-disk of an old computer. This can be expressed as $(size)^{C/size}$ where C is a proper constant. As the file size increases, the exponent $C/size$ decreases so that the number of files of large size is far less than smaller ones. The PDF is sometimes expressed in size-modeling studies using a variable s as $f(s; k, c) = c k^c / s^{(c+1)}$.

Processes that exhibit power laws follow a random multiplicative growth. Thus, the ratio S_{t+1}/S_t, called the growth factor or growth rate,[3] is independent of the current size S_t. This means that the parameter k can be interpreted in two different ways: either as the leftmost variate value or as a compounded measure of the observed growth rate. The tail index is reparameterized as $c = 1/\zeta$ in some application areas.

This distribution is related to exponential distribution as follows: If Y is exponentially distributed with PDF $f(y) = \exp(-y)$, then $X = k \exp(Y/c)$ is distributed as Pareto(k, c) given in Equation (12.1). In other words, a log-transformed Pareto variate has an exponential distribution. Similarly, the stopping time of a Brownian motion when assumed to be exponentially distributed

---

[2]Some authors call k as the location parameter, which is incorrect because location refers to the center of gravity which may be near k due to the reverse J-shape, but actually depends on the other parameter. A location parameter m can be introduced to get the CDF as $F(x; m, k, c) = 1 - ((x - m)/k)^{-c}$.

[3]Called Gibrat's law of proportional growth in economics.

results in a double Pareto distribution (which is a mixture of two Pareto distributions). Personal income models in finance uses the multi-parameter Champernowne distribution, the simplest of which with one parameter has CDF $F(x;a) = 1 - 1/(1 + x^a)$ for $x > 0$. The zipf distribution (also called zeta distribution) is the discrete analogue of Pareto distribution. The Benini distribution with CDF $F(x;a,b,c) = 1 - (x/c)^{-a+b\log(x)}$ for $x \geq c$ is a modification for $b \neq 0$, and reduces to Pareto distribution for $b = 0$.

As $c \to \infty$, the PDF approaches Dirac's δ function. Left truncation and size-biasing results in Pareto distributions itself. The sum of the logarithm of several independent scaled Pareto distributions has a gamma distribution, whereas the limiting distribution of generalized gamma distribution is Pareto distributed (James (1979) [98]). If X_j for $j = 1, 2, \ldots, n$ are IID Par-I(k, c) random variables, then $Y = 2c \ln(\prod_{j=1}^{n} X_j/k^n)$ has a central χ^2 distribution with $2n$ degrees of freedom (DoF). It is also related to the Beta-I (a, b) distribution.

There exist several practical problems in which the empirical distribution is J-shaped. Examples include the storage capacity of personal computers (from 1980s to the present), speed of CPU, number of pages on the World Wide Web (WWW), number of vehicles in a city, etc. A simple transformation $Y = 1/X$ brings this into the Pareto form so that the techniques discussed below can be used to fit such models.

Problem 12.1   If X$\sim$ Pareto(a, k), find the distribution of $Y = 1/X$, and show that the moments of Y are related to the inverse moments of X. What are the conditions for the existence of the moments of Y?

Problem 12.2   If the parameter θ of an exponential distribution $f(x, \theta) = (1/\theta)\exp(-x/\theta)$ is gamma distributed, prove that the unconditional distribution has Pareto distribution.

Problem 12.3   A dental insurance company observes that the number of claims has a geometric distribution with parameter θ. If the prior distribution of θ is Pareto-I(1,k), find the unconditional distribution.

12.3   PROPERTIES OF PARETO DISTRIBUTION

This distribution has a slowly decaying tail to the right (called fat-tailed or heavy-tailed law[4]), which agrees perfectly with several natural phenomena like wealth distribution, insurance claims, species abundance in biology, earthquake magnitudes, fault sizes at certain locations and energy releases from seismic activities, area destroyed by forest fires, length of words in different natural languages, etc. Hence, it is used to model "bigger is rare" situations (large ones are rare, but important; and small ones are common and unimportant). It is also used as the "distribution of error-terms" (instead of the standard assumption of Gaussian error distribution) in regression and time-series fitting of skewed financial data. The support is dependent upon the scale parameter k. As the distribution is reverse

---

[4]A fat-tailed distribution has a survival function that decays as a power law.

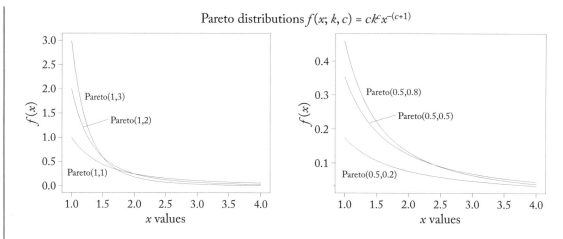

Figure 12.1: Pareto distributions.

J-shaped, the mode is k, and the axes are the asymptotes. The integral and derivative of the PDF have similar forms. The log-log plot of this distribution is linear with a negative slope because taking log of $f(x; k, c) = ck^c x^{-(c+1)}$ gives $\log(f(x)) = K - (c + 1)\log(x)$ where K is a constant, so that the absolute value of the slope can be used as a rough estimate of the unknown c, and the antilogarithm ($\exp()$) of the ratio $|intercept/slope|$ as a rough estimate of k. Pareto postulated that the number of persons in a population with wealth or income $\geq x$ can be approximated by the SF given below, although in practice this holds only for restricted values of x, usually beyond the mode. This came to be known as "Pareto tail behavior."

The CDF of $f(x; c) = cx^{-(c+1)}$ is $F(x; c) = 1 - x^{-c}$, and that of $f(x; k, c) = ck^c x^{-(c+1)}$ is $F(x; k, c) = 1 - (x/k)^{-c}$. A literal meaning of this is that the fraction of data values (or the probability associated with it) whose size exceeds a fixed threshold K decays like a power function. As $f'(x) < 0 \forall x$ and $(\ln(f(x)))'' = (c + 1)/x^2$, the density function is log-convex whereas the CDF is log-concave. The inverse distribution function is $F^{-1}(p) = k/q^{1/c}$ where $q = 1 - p$. SF of this distribution takes the simple form $S(x) = (k/x)^c$, so that $\log(S(x)) = c[\log(k) - \log(x)]$. This suggests that the distribution is closed under the *minima* operation among a set of finite IID Pareto random variables (Balakrishnan and Nevzorov (2003) [15]). In those applications in which time is the variable of interest, define $\bar{T}(x) = 1/S(x)$, that denotes the probability of occurrence of the event of magnitude smaller than x.

The CV is $1/\sqrt{c(c - 2)}$, for $c > 2$, which is independent of k. As the CDF has a slightly simpler form, it is more suitable for extensions and characterizations. Thus, the Pareto distribution of second kind (Par-2(a,C) or Pareto-II(a,C)) has CDF $F(x; a, C) = 1 - C^a/(C + x)^a = 1 - [C/(C + x)]^a$ for $x \geq 0$. This can also be written as $F(x; a, C) = 1 - 1/(1 + x/C)^a$, for which the quantiles have a closed form as $F^{-1}(p) = C((1 - p)^{-1/a} - 1)$ for $0 < p < 1$.

Pareto distribution of third kind (Par-3(a,b,C) or Pareto-III(a,b,C)) has CDF $F(x; a, b, C) = 1 - C \exp(-bx)/(C + x)^a$ for $x \geq 0$.

Problem 12.4   An insurer's annual loss follows a Pareto distribution with PDF $f(x) = (5/2)(300)^{5/2}/x^{7/2}$, for $x \geq 300$. Find the median loss and inter-quartile range $Q_3 - Q_1$.

12.3.1   MOMENTS AND GENERATING FUNCTIONS

Raw moments are easy to find due to the presence of x in the denominator. However, the mean and other moments may not always exist. This is because of the fact that when the shape parameter c is very small, the integral for the moments $E(X^k)$ may not converge. Thus, the mean μ exists only when $c > 1$, and the variance exists only when $c > 2$.

Example 12.5   Moments of Pareto distribution   Prove that the rth moment of Pareto distribution is $\mu_r = c * k^r/(c - r)$ for $r < c$.

Solution: As the range of x is from k to ∞, we have

$$\mu'_r = ck^c \int_k^\infty x^r/x^{c+1}dx = ck^c \int_k^\infty x^{r-c-1}dx = ck^c/(r-c)x^{r-c}||_k^\infty. \qquad (12.6)$$

As the integrand is a power of x, it converges for $r < c$ to get $ck^c/(r - c)(0 - k^{r-c})$. The k^c cancels out giving $\mu_r = c * k^r/(c - r)$. Take c as a common factor from denominator to get $\mu_r = k^r(1 - r/c)^{-1}$. This gives the recurrence relation $\mu_{r+1} = \mu_r * k * (c - r)/(c - r - 1)$. The mean and variance of the standard form $f(x; c) = cx^{-(c+1)}$ are $c/(c - 1)$ and $c/[(c - 1)^2(c - 2)]$, from which by equating the mean, we get an estimate of c as $\hat{c} = \overline{x}/(\overline{x} - 1)$. From this the mean and variance are easily obtained as $\mu = kc/(c - 1), \sigma^2 = k^2c/[(c - 1)^2(c - 2)]$. Thus, the mean μ is infinite, or undefined when the shape is ≤ 1. The mean and variance for the alternate representation $f(x; k, c) = (c/k)(1 + x/k)^{-(c+1)}$ are $k/(c - 1)$ and $ck^2/[(c - 1)^2(c - 2)]$.

The rth raw moment of Pareto-II(a,C) is $E(X^r) = r!C^r/[(a - 1)(a - 2)\ldots(a - r)]$ for $a > r$, from which the mean and variance are respectively $\mu = C/(a - 1)$ for $a > 1$, and $\sigma^2 = aC^2/[(a - 1)^2(a - 2)]$ for $a > 2$. As $\sigma^2/\mu = aC/[(a - 1)(a - 2)] = \mu a/(a - 2)$, the variance is quite often greater than the mean. This reduces to the standard Par-2 distribution for $C = 1$. Median is $k * 2^{1/c}$.

Example 12.6   Parameters of Pareto distribution   An insurance company found that the claim size for a particular type of insurance is Par-II(60, 4225). Find the unknown parameters and 90th percentile of the distribution

Solution: Equate the mean to get $C/(a - 1) = 60$. Similarly, equating variance gives $\sigma^2 = aC^2/[(a - 1)^2(a - 2)] = 4225$. Square the first equation and put in the second one to get $4225 =$

$(60)^2 a/(a-2)$ or $1 - 2/a = 3600/4225 = 0.852071$. From this $a = 6.760 * 2 = 13.52$, and $C = 751.2$. The 90th percentile satisfies $1 - (751.2/(\alpha + 751.2)) = 0.90$.

Problem 12.7　If the exceedance of a dam is Pareto distributed with mean and variance 50 and 1600, find the parameters of the distribution.

Problem 12.8　Annual claims in a dental insurance plan has a Par-I(150,c) distribution where c is unknown. Estimate the unknown if the claims for three policies are 150, 420, 630.

Example 12.9　Mean deviation of Pareto distribution
　　Find the mean deviation of the Pareto distribution $f(x, k, c) = ck^c x^{-(c+1)}$ where $x \geq k$.

Solution:　Using the power method in Chapter 1 of Chattamvelli and Shanmugam (2021) [47] we get the MD as

$$MD = 2 \int_{ll}^{\mu} F(x)dx = 2 \int_{k}^{d} [1 - (k/x)^c]dx, \quad \text{where } d = kc/(c-1). \tag{12.7}$$

Separate into two terms and integrate each term to get $MD = 2\{[kc/(c-1) - k] - k^c \int_k^d x^{-c} dx\}$. The expression inside the square bracket simplifies to $k/(c-1)$, and the integral simplifies to $(k/(1-c))(c-1)^{c-1}/c^{c-1} + k/(1-c)$. The term $k/(1-c)$ cancels with first term $k/(c-1)$. Take c outside from the second expression to get $MD = 2k(1 - 1/c)^{c-1}/(c-1)$. See Table 12.1 for further properties.

12.3.2　FITTING

If k is known, equating the population mean $\mu = kc/(c-1)$ and the sample mean $\bar{x}$ gives an estimate of c as $\bar{x}/(\bar{x} - k^*)$. Let $x_1, x_2, \ldots, x_n$ be a random sample of size n from a Pareto population. Due to the reverse-J shape, an intuitive estimate of k is $\hat{k} = x_{(1)}$, the smallest value (first order-statistic). If $x_{(1)}$ is the first order statistic in a sample of size $n > 1$, Quandt (1966) [179] obtained a consistent estimator of c as $(n\bar{x} - x_{(1)})/[n(\bar{x} - x_{(1)})]$. If k is known, the log-likelihood function is given by

$$l(c) = n \log(c) + nc \log(k) - (c+1) \sum_{j=1}^{n} \log(x_j). \tag{12.8}$$

Differentiate wrt c and equate to zero to get $\partial l(c)/\partial c = n/c + n \log(k) - \sum_{j=1}^{n} \log(x_j) = 0$. This gives $\hat{c} = 1/[\sum_{j=1}^{n} \log(x_j) - n \log(k)]/n = 1/[\frac{1}{n} \sum_{j=1}^{n} \log(x_j) - \log(k)] = 1/\log(GM/\hat{k})$. Shanmugam (1987) [200] showed that the MLE is insufficient compared to MVUE (Minimum Variance Unbiased Estimate) when the parameter c is known. Bayesian estimation approach uses gamma law as natural conjugates for c and power-law for k. Using a characterization of Pareto distribution, Shanmugam (1999) [201] derived a statistic to test if a sample has come from a Pareto population.

Table 12.1: Properties of Pareto distribution ($f(x; k, c) = ck^c x^{-(c+1)}$)

Property	Expression	Comments
Range of X	$k < x < \infty$	Continuous
Mean	$\mu = kc/(c - 1)$	$k[1 + 1/(c - 1)]$
Median	$k2^{1/c}$	Mode = k
Variance	$\sigma^2 = k^2 c/[(c - 1)^2(c - 2)]$	$= \mu^2/[c(c - 2)]$
Skewness	$\gamma_1 = 2[(c - 2)/c]^{1/2}(1 + c)/(c - 3)$	Valid for $c > 3$
Kurtosis	$\beta_2 = 6(c^3 + c^2 - 6c - 2)/[c(c - 3)(c - 4)]$	Valid for $c > 4$
Mean deviation	$E\lvert X - \mu \rvert = 2k(1 - c^{-1})^{c-1}/(c - 1)$	
CV	$1/\sqrt{c(c - 2)}$	$c > 2$
SF	$(k/x)^c$	
CDF	$1 - (k/x)^c$	
Moments	$\mu_r = c * k^r/(c - r)$	
MGF	$c(-kt)^c \Gamma(-c, -kt)$	For $t < 0$
ChF	$c(-ikt)^c \Gamma(-c, -ikt)$	

Put $x = y - b$ to get a three-parameter version. Note that the expression for skewness is valid for $c > 3$, and it is never symmetric (see Figure 12.1). β_2 given here is the excess kurtosis, $\Gamma()$ is gamma function.

12.3.3   RANDOM NUMBERS

As the CDF is $F(x) = 1 - (k/x)^c$, it is easy to generate random numbers using the inverse-CDF method. Let u be a uniform random number (in $(0, 1)$ range). Equate $u = 1 - (k/x)^c$ and solve for x to get $(1 - u)^{1/c} = k/x$ or $x = k/(1 - u)^{1/c}$. Take log of both sides to get $y = \log(x) = K - C \log(1 - u)$ where $K = \log(k)$ and $C = 1/c$. For $n > 1$, generate n random numbers $(u_1, u_2, \ldots, u_n) \in (0, 1)$, and find y_i using the above relationship. Exponentiate them to get $x_i = \exp(y_i)$.

12.4   EXTENSIONS OF PARETO DISTRIBUTION

Several generalizations of the distribution are available. A two-parameter generalized Pareto distribution (GPD) has CDF

$$F(x; a, b) = \begin{cases} 1 - (1 + ax/b)^{-1/a} & \text{if } a \neq 0; \\ 1 - \exp(-x/b) & \text{if } a = 0, \end{cases}$$

where $b > 0$ and $x \geq 0$ for $a > 0$, and $0 \leq x \leq -b/a$ otherwise. Another two-parameter GPD with CDF

$$F(x; a, b) = 1 - (1 - ax/b)^{1/a}, \tag{12.9}$$

which reduces to the exponential distribution $\text{EXP}(1/b)$ for $a \to 0$, the standard Pareto distribution for $a < 0$ and Pareto-II for $a > 0$. A location parameter can be introduced in the above to get a three-parameter GPD. A four-parameter GPD has CDF

$$F(x; a, b, c, d) = 1 - \left(\frac{c + d}{c + x}\right)^a \exp(-b(x - d)), \quad x \geq d. \tag{12.10}$$

The GPD distributions are used to model "excess of data values over a fixed threshold," although any continuous heavy-tailed distribution can be used for this purpose. Consider a continuous distribution with CDF $F(x)$. Then the excess distribution over a threshold c has CDF

$$F(x; c) = Pr[X - c \leq x | X > c] = (F(x + c) - F(c))/(1 - F(c)), \tag{12.11}$$

which is called "residual life" CDF in survival analysis. This reduces to the usual CDF for exponential distributions, reflecting its memory-less property. The mean excess function for finite mean can be defined as $E[X - c | x > c]$.

Problem 12.10   Find $F(x|x > a = F(x)/(1 - F(a)), \quad a < x < \infty$ for the Pareto distribution.

The GPD arises when the threshold is allowed to vary toward the right tail, instead of keeping it fixed. This usually happens in mechanical engineering systems that undergo vibrations, rotating machinery or parts due to wear and tear, and volatility in financial markets. A symmetric Pareto distribution can be defined by replacing x by $|x|$, changing the range as $|x| > k$, and adjusting the normalizing constant.

12.4.1   TRUNCATED PARETO DISTRIBUTIONS

Due to the limitation of variables of interest, most of the empirical studies use a right-truncated Pareto distribution with PDF

$$f(x; k, c, r) = (c/k)(k/x)^{c+1} [1 - (k/r)^c]^{-1}, \quad k \leq x \leq r. \tag{12.12}$$

Problem 12.11   Find the mean and variance of truncated Pareto distribution.

12.4.2   SIZE-BIASED PARETO DISTRIBUTIONS

As the mean is $kc/(c - 1)$, a size-biased Pareto distribution can be obtained as $f(x; k, c) = ck^c x^{-(c+1)} * x(c - 1)/kc$. This simplifies to $f(x; k, c) = (c - 1)k^{c-1}x^{-c}$, which is again Pareto distributed. Alternately, find the expected value of $(1 + mx)$ as $E(1 + mx) = 1 + mkc/(c - 1)$ and proceed as discussed in Chapter 1 of Chattamvelli and Shanmugam (2021) [47]. Many other generalizations of Pareto distribution exists. See, for example, Kumaraswamy Pareto distribution (Bourguingon, et al. (2013) [26]), exponentiated Pareto distribution (Nadarajah (2005) [160]), and the references therein. A type-II Pareto distribution has PDF

$$f(x; a, c) = (a/c)(1 + x/c)^{-(a+1)}, \quad a, c, x > 0. \tag{12.13}$$

Weighted versions of (12.13) (say with weight x^k) is used in ecology and reliability engineering, the CDF of which can be expressed as

$$F(x; a, k, c) = \Gamma(a + 1)/[c^k \Gamma(a - k)\Gamma(k + 1)] \, I(x/c; k + 1, a - k), \quad a, c, x > 0, \tag{12.14}$$

where $I(x; a, b)$ is the incomplete beta function.

12.5 APPLICATIONS

The Pareto distribution is used to model socio-economic and other naturally occurring quantities such as wealth and income data, operational risk losses in business, employee headcount of firms in business, claimed amounts and losses in actuarial sciences, exceedences in hydrology, and in reliability, and life testing experiments ([186], [200, 201]). As examples, the "family names" (also called lastname or surname)[5] in some countries, consumption of non-essential commodities (like alcohol, some soft drinks) exhibit the Pareto distribution. The probability distribution of the back-scatter of fully polarimetric X-band radars used in sea surface surveillance uses Pareto distribution because it can model the high-magnitude components of the sea-clutter and work well with the thermal noise generated in the radar (Rosenberg and Bocquet (2015) [186]). It is also used to approximate the voltage distribution of isolated-neutral capacitance networks during the impact of surge in electronics engineering, anisotropic boundary diffusion at tilt boundaries in thermal engineering. It works best in those situations where the frequencies tail off to the right very slowly. Astronomers use it to model the size of solar flares, crater sizes on other planets and moons, etc.

Analyzing extreme tails can have a major macro-economic impact in those fields where the variable of interest is money (tax payments, collections from movies, insurance claims, and mutual funds). Although the Pareto distribution may provide a better fit for the extreme tails, it can be a poor fit for the entire data. Alternative choices in such situations are the lognormal, skew-normal and extreme value distributions. For instance, the load profile of electric load in low-voltage grids is usually modeled using log-normal law. However, due to the rapid rise in e-mobility and scarcity of public charging stations, many customers resort to domestic recharging of their e-vehicles. The load demand in such situations is modeled using GPD, which gives a better fit (Probst, Braun, Backes, and Tenbohlen (2011) [176]). Although other distributions give better overall fit, Pareto distribution and its extensions are the preferred choice in extreme tail modeling (say the top 5%) as in firm size and labor income modeling, aquatic ecology, oceanography, marine geology, and earth sciences.

[5]See Fox and Lasker (1983) [71] for the distribution of surname frequencies.

Another technique has arisen recently by dividing the entire distribution range into a "bulk-part" and the remaining "tail part," and fitting entirely different distributions to each part. The standard practice is to fit the tail part using the GPD separately. This is known as the peaks-over-threshold (POT) in extreme value theory. This outperforms other single distribution models in many fields. For instance, Yakovenko and Silva (2005) [234] uses an exponential bulk and power-law tail for income distributions in the U.S. The intersection of these two curves determine the income separating the two classes. Economists use the Lorentz curve to study the income inequalities. This is obtained by plotting the fraction of the population with income less than c along the X-axis and the total income of this population along the Y-axis. The following paragraphs list some of the more popular applications.

12.5.1   INSURANCE

The Pareto distribution has been a popular choice in actuarial sciences. It is used to model high-risk insurance claims and severity of large casualty losses. Claimed amounts can be on the higher side in some insurance fields (like fire and health insurance). It is the usual practice to model losses in excess of a threshold in extreme insurance claims. GPD is one choice in such domains. Alternatively, the tail of the severity distribution is modeled using Pareto law and the bulk by another law.

12.5.2   INVENTORY CONTROL

Several industries like biomedical, pharmaceutical, and food-processing work with perishable items in their inventory. These usually have a preset lifetime under controlled storage conditions. Items with long lifetime (slow moving) are modeled as Pareto decay. A single Pareto distribution may not be appropriate to model the lifetime in all situations. One simple approach is to use a convex combination of two models, the simplest of which is two Pareto models Par-I(k, c_1) and Par-I(k, c_2) with respective weights α and $(1 - \alpha)$.

12.6   SUMMARY

The Pareto distribution has its inception in finance and taxation. It has been applied for more than a century in many other fields mentioned above. The primary reason for the popularity of Pareto distribution is its simplicity. A new trend in some fields is to model the extreme tails using the Pareto law and the bulk part by another law like the IGD, LND, or Weibull model.

CHAPTER 13

Laplace Distribution

13.1 INTRODUCTION

This distribution, introduced by Pierre–Simon Laplace (1749–1827) in 1774, has many applications in geology, hydrology and ocean engineering, economics and finance, quality control, image and speech recognition, inventory control (especially of slow moving items), scientometrics, pharmacokinetics (drug absorption, delivery, depletion from the body), nuclear medicine, telecommunications, and error modeling (called Laplacian noise) in various fields like astronomy, global navigation systems, etc. Also called the *first law of Laplace, two-tailed exponential, bilateral exponential distribution*, or *double-exponential (DE) distribution* (because each part on the left and right of the location parameter is proportional to an exponential density or its mirror image), it is also used in life testing models with heavy tails, and in machine learning (natural language processing, computer vision, and intrusion detection). The PDF of classical Laplace distribution (CLD) with two-parameters is given by $f(x; a, b) =$

$$(1/(2b)) \exp(-|x-a|/b), \ = (0.50/b) \exp(-|x-a|/b), \ -\infty < a, x < \infty, \ b > 0, \quad (13.1)$$

where "a" is the location parameter (mean) and "b" is the scale parameter (also called diversity). It is denoted as $\mathfrak{L}(a, b)$, $Lp(a, b)$, or Laplace(a, b). This can also be written as

$$f(x; a, b) = \frac{1}{2b} \begin{cases} \exp(-(a-x)/b) & \text{if } x < a; \\ \exp(-(x-a)/b) & \text{otherwise.} \end{cases}$$

It is symmetric around a, and the right part resembles an exponential or Pareto distribution. The standard Laplace distribution (SLD) is obtained by putting $a = 0, b = 1$, with PDF $f(z) = (1/2)e^{-|z|}$; and the zero-mean Laplace distribution by putting $a = 0$ with PDF $f(x; 0, b) = (1/2b) \exp(-|x|/b)$. A one-parameter (location) unit variance Laplace distribution has PDF $f(x; a, 1) = \frac{1}{\sqrt{2}} \exp(-\sqrt{2}|x-a|)$. As the exponent involves absolute values, it is appropriate in the study of first order phenomena like low-dose response modeling in medical sciences (where the log-Laplace distribution discussed below is preferred). Put $1/b = 2c$ to get another form $f(x; a, c) = c \exp(-2c|x-a|)$, $1/b = c$ to get $f(x; a, c) = (c/2) \exp(-c|x-a|)$, and $\sqrt{2}b = c$ to get $f(x; a, c) = 1/(\sqrt{2}c) \exp(-\sqrt{2}|x-a|/c)$ (which is the preferred form in astronomical sciences).

13.2   RELATION TO OTHER DISTRIBUTIONS

If $X \sim$ Laplace(a, b), then a change of origin and scale transformation $cX + d$ results in Laplace$(ac + d, bc)$. The transformation $Y = |X - a|/b$ results in the standard exponential distribution. Similarly, $Z = (2/b)|X - a|$ has a χ^2 distribution with 2 DoF. The standard Laplace distribution is the distribution of the difference of two IID standard exponential variates. In general, if X_k, $k = 1, 2$ are two independent EXP(λ) variates, then $Y = X_1 - X_2$ has a Laplace $(0, \lambda)$ distribution. Conversely, if $X \sim$ Laplace$(0, \lambda)$ then $Y = |X| \sim$ EXP$(1/\lambda)$. If X_k for $k = 1, 2, \ldots, n$ are IID Laplace(a, b) random variables, $Z = (2/b) \sum_{k=1}^{n} |X_k - a|$ has a χ^2_{2n} distribution with $2n$ DoF. From this it follows that the ratio of two IID Laplace variates Laplace$(0, b_1)$ and Laplace$(0, b_2)$ is a central F$(2, 2)$ variate. If $X \sim$ Laplace$(0, b)$ then $Y = |X|$ has an EXP$(1/b)$ distribution. If $X \sim$ EXP(1) and $Z \sim$ N$(0, 1)$ are IID, then $Y = a + bZ\sqrt{2X}$ is distributed as Laplace(a, b). If $X \sim$ N$(0, 1)$ is independent of Y which is distributed as Rayleigh with PDF $f(y) = y \exp(-y^2/2)$, then $W = XY$ is an SLD. If U_1, U_2 are IID standard uniform ($U(0, 1)$) random variables, then $Y = \ln(U_1/U_2)$ is an SLD. This property can be used to generate random numbers from Laplace distribution as U and $1 - U$ are identically distributed. Laplace and Cauchy distributions are related through characteristic functions, as shown below. If Z_1, Z_2, Z_3, Z_4 are IID N$(0, 1)$ random variables, $Z_1 Z_2 \pm Z_3 Z_4$ is Laplace$(0, 2)$. The asymmetric Laplace distribution has PDF

$$f(x; a, b, c) = \frac{1}{c} \frac{b}{1 + b^2} \begin{cases} \exp(-c|x - a|/b) & \text{for } x < a; \\ \exp(-bc|x - a|) & \text{for } x \geq a. \end{cases}$$

(Jones (2002, 2019) [106, 110]). The ratio of two IID classical Laplace distributions is called the double Lomax distribution (DLD). DLD can be obtained by compounding CLD with an exponential density. Discrete version of SLD has been obtained by Kozubowski and Inusah (2006) [120] with PMF

$$f(x; p) = (1 - p)/(1 + p)p^{|k|}, \quad k = 0, \mp 1, \mp 2, \ldots; 0 < p < 1, \tag{13.2}$$

which arises as the difference of two IID geometric random variables.

13.3   PROPERTIES OF LAPLACE DISTRIBUTION

This distribution is symmetric around $x = a$, for which the sample median is the MLE. It has a discontinuous first derivative at the location parameter. Put $1/b = \lambda$ to get the form

$$f_x(a, \lambda) = (\lambda/2) \exp(-\lambda|x - a|), \quad -\infty < x < \infty, \quad -\infty < a < \infty, \quad \lambda > 0. \tag{13.3}$$

Take log to get $\ln(f(x)) = C - \lambda|x - a|$, which is a concave function in x, although the derivative is nonexistent at $x = a$.

Due to symmetry, β_1 and all odd moments except the first are zeros. The mean, median, and mode are $\mu = a$, and the variance is $\sigma^2 = 2b^2$. The modal value is $1/(2b)$, showing that the

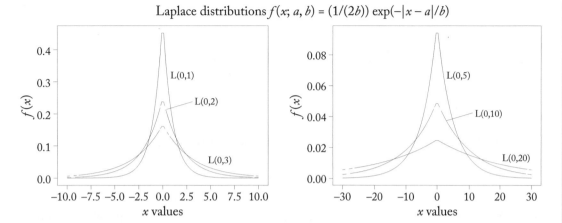

Figure 13.1: Laplace distributions.

peakedness increases rapidly as $b \to 0$ (Figure 13.1). This is considered to be an anomaly in some fields like stock markets and finance. Whereas the density of the arcsine distribution is peaked at the extreme (boundary) values, that of Laplace distribution is peaked at the location value. In this sense, this distribution is just the opposite of arcsine (or Beta-I$(1/2, 1/2)$ on $[0, 1]$) distribution. When $a = 0$, the resulting distribution and that of two IID exponential distributions are the same, and the PDF on $x > 0$ equals half the probability of the exponential distribution.

As the absolute value is used in the exponent, this distribution tails-off slower than the normal law for which the exponent part is half the squared deviation from the mean ($\exp(-\frac{1}{2}((x - \mu)/\sigma)^2)$). See Figure 13.1. Geometrically, it has steeper peaks and heavier tails than the Gaussian law. Whereas the Fourier coefficients of Gaussian distribution decay exponentially, those of Laplace distribution decay polynomially. As several statistical inferences involve tail probabilities of hypothesized distribution, it is important to choose the most appropriate one when working with empirical data. The quantity $E(\hat{p}(x)) = \int_{x-0.5}^{x+0.5} f(x)dx$, called expected bias estimate, is used in some engineering fields. This becomes $\frac{1}{2}\int_{x-0.5}^{x+0.5} \exp -|z|dz = -\frac{1}{2}\exp(-z)|_{x-0.5}^{x+0.5} = \frac{1}{2}(\sqrt{e} - 1/\sqrt{e}) = 0.521095$ for the SLD. Similarly, $\Pr[a \leq X \leq a + \delta] = (1/2)\int_a^{a+\delta} \exp(-|z|)dz$. This is approximately $\delta f(a)$. Although the tails are heavier than normal tails, it is lighter than that of Cauchy law due to the steeper peak. An implication of this is that all moments of SLD exist while they are nonexistent for the Cauchy law.

Example 13.1 Even moments of Double Exponential Prove that the even moments of double exponential are given by $\mu_{2k} = (2k)!b^{2k}$.

Solution: All odd moments are zeros as the distribution is symmetric. The even moment is given by $\mu_{2k} =$

$$\int_{-\infty}^{\infty} (x-\mu)^{2k} \frac{1}{2b} \exp(-|x-a|/b)\, dx = \frac{1}{2b}[\int_{-\infty}^{\mu} (x-\mu)^{2k} \exp(-|x-a|/b)\, dx$$
$$+ \int_{\mu}^{\infty} (x-\mu)^{2k} \exp(-|x-a|/b)\, dx]. \quad (13.4)$$

As $\mu = a$, put $y = (x-a)$ and change the limits accordingly to get

$$\mu_{2k} = \frac{1}{2b}[\int_{-\infty}^{0} y^{2k} \exp(y/b)\, dy + \int_{0}^{\infty} y^{2k} \exp(-y/b)\, dy]. \quad (13.5)$$

Put $-y = t$ in the first integral. Then it becomes the second integral. Hence, $\mu_{2k} = 2\int_{0}^{\infty} y^{2k} \exp(-y/b)\, dy$. Write y^{2k} as $y^{(2k+1)-1}$. Using gamma integral this becomes $\Gamma(2k+1)b^{2k}$. As $\Gamma(2k+1) = (2k)!$, the result follows. The absolute central moments are also easy to find as

$$\nu_k = E|X-a|^k = \frac{1}{2b}[\int_{-\infty}^{0} (a-y)^k \exp(y/b)\, dy + \int_{0}^{\infty} (y-a)^k \exp(-y/b)\, dy]. \quad (13.6)$$

Using gamma integral this becomes $b^a \Gamma(a+1)$. Integrate wrt x to get the CDF as

$$F_x(a,b) = (1/2)[1 + \text{sign}(x-a)(1 - \exp(-|x-a|/b)], \quad -\infty < x < \infty, \quad (13.7)$$

from which $F(a) = 1/2$, and $F(0) = (1/2)[1 - (1 - \exp(-a/b))] = (1/2)\exp(-a/b)$, for $a \neq 0$. From this the inverse CDF follows as

$$x = \begin{cases} b \ln F(x) + a & \text{if } F(x) < 0.50; \\ -b \ln(2(1 - F(x))) + a & \text{otherwise.} \end{cases}$$

The CDF of SLD can be written as

$$F(x) = \begin{cases} \exp(x)/2 & \text{if } x < 0; \\ 1 - \exp(-x)/2 & \text{otherwise.} \end{cases}$$

These can be combined and written as

$$F(x) = \frac{1}{2}\left(\text{sign}(x)[1 - \exp(-|x|)]\right).$$

13.3.1   GENERATING FUNCTIONS

The characteristic function is obtained easily as follows:

$$\phi(t) = \frac{1}{2b} \int_{-\infty}^{\infty} e^{itx} e^{(-|x-a|/b)}\, dx. \quad (13.8)$$

Split the range of integration from $-\infty$ to a and from a to ∞, and change the variable as $y = (x - a)/b$ to get

$$\phi(t) = \frac{e^{iat}}{1 + b^2 t^2}, \qquad (13.9)$$

from which the MGF is $M_x(t; a, b) = \exp(at)/(1 - b^2 t^2)$, and ChF is $\phi_x(t; a, b) = \exp(iat)/(1 + b^2 t^2)$ for $-1/b < t < 1/b$. Extensions of this distribution can be obtained by introducing additional parameters in the ChF. For example, $\phi_x(t; a, b, c) = \exp(iat)/(1 + b^2 t^2)^c$ and $\phi_x(t; a, b, c, d) = \exp(iat)/(1 - i\,dt + b^2 t^2)^c$ result in a new distributions of the same type. The mean and variance can be obtained easily as $\mu = a$ and $\sigma^2 = 2b^2$, showing that this belongs to the location-and-scale family of distributions. As $\sigma = \sqrt{2}b$, the scale parameter is the population standard deviation divided by $\sqrt{2}$. The CV is $\sqrt{2}b/|a|$, and the kurtosis is 6 (which is double the kurtosis of standard normal, so that it belongs to the "super Gaussian" distributions). If Y is the sum of n IID Laplace$(0, b)$ random variables, the distribution of it can be obtained using ChF as

$$f(y) = \sum_{k=0}^{n-1} \binom{n + k - 1}{k} b^{n-k}/[2^{n-k}\Gamma(n - k)] \exp(-b|y|)|y|^{n-k-1}, \qquad (13.10)$$

which is a weighted sum of double gamma distributions. For $a = 0$, $b = \pm 1$ the MGF becomes $\phi(t) = 1/(1 + t^2)$, which shows that Laplace and Cauchy distributions are related through characteristic functions. By decomposing $1/(1 + t^2) = \frac{1}{2}\left(\frac{1}{1-it} + \frac{1}{1+it}\right)$, it follows that the SLD is a convex combination of exponential mixture with weights 1/2 each. The MGF and CGF of SLD follow from this as $M_x(t) = 1/(1 - t^2)$, and $K_x(t) = -\log(1 - t^2)$. Expand $-\log(1 - t^2)$ as an infinite series to get $K_x(t) = t^2 + t^4/2 + t^6/3 + \cdots + t^{2r}/r + \cdots = t^2(1 + t^2/2 + t^4/3 + \cdots + t^{2(r-1)}/r + \cdots)$, which is obviously an even function. Multiply numerator and denominator by $(2r)!$ to get the $(2r)^{th}$ cumulant as $k_{2r} = (2r)!/r$. As the last term in the expansion of the factorial $(2r)!$ is $2r$, which when divided by r gives a 2, we get the cumulant as $k_{2r} = 2(2r - 1)!$. Substitute $2r = s$ to get $k_s = 2(s - 1)!$ where s is even. The first few cumulants are $k_2 = 2$, $k_4 = 2 \times 3! = 12$, $k_6 = 2 \times 5! = 240$, etc. The cumulants of SLD satisfy the recurrence $k_{2r+2} = 2r(2r + 1)k_{2r}$. CGF in the general case is $K_x(t) = at - \log(1 - b^2 t^2)$. Proceed as above to get the cumulants as $k_1 = a$, $k_2 = 2b^2$, $k_4 = 12b^3$, $k_{2r} = (2r)!b^{2r}/r = 2b^{2r}(2r - 1)!$. These satisfy the recurrence $k_{2r+2} = 2b^2\, r(2r + 1)k_{2r}$.

The quartiles are easy to find from the CDF as

$$x_p = \begin{cases} a + b\log(2p) & \text{if } p \in (0, 0.50]; \\ a - b\log(2q) & \text{for } q = 1 - p, \text{ if } p \in (0.50, 1). \end{cases}$$

In particular, $Q_1 = a - b\log(2)$ and $Q_3 = a + b\log(2)$. In some applications, a random noise may have to be added to the observed data values to mimic a known process. The Gaussian noise is used when the tails are narrow and Laplacian noise otherwise. The distribution of the median of a sample of size n is $f(x) = C\exp((n + 1)\sqrt{2}x)(1 - \exp(\sqrt{2}x)/2)^n$ where $C = (n + 1)\binom{2n+1}{n}/2^{n+1/2}$ (Lawrence (2013) [126]). See Table 13.1 for further properties.

Table 13.1: Properties of Laplace distribution

Property	Expression	Comments
Range of X	$-\infty < x < \infty$	Continuous
Mean	$\mu = a$	
Median	a	Mode $= a$
Variance	$\sigma^2 = 2b^2$	
Skewness	$\gamma_1 = 0$	Always symmetric
Kurtosis	$\beta_2 = 6$	
Mean deviation	$E\lvert X - \mu\rvert = b$	
CV	$\sqrt{2}b/\lvert a\rvert$	
CDF	$\frac{1}{2}\exp(-(a-x)/b)$ $\quad x < a$ $1 - \frac{1}{2}\exp(-(x-a)/b)$ $\quad x > a$	
Moments	$\mu_r = \Gamma(r+1)b^r$	r even
MGF	$\exp(at)/(1 - b^2t^2)$	$\lvert t\rvert < 1/b$
ChF	$\exp(iat)/(1 + b^2t^2)$	
Entropy	$1 + \ln(2b)$	

Problem 13.2   Show that the mean and variance of $f(x; a, b, c) = (ab/2)(1 + b\lvert x - c\rvert)^{-(a+1)}$ are $\mu = b/(a-1)$ and $\sigma^2 = 2b^2/[(a-1)(a-2)]$.

Example 13.3   Mean deviation of Laplace distribution   Find the mean deviation of the Laplace distribution.

Solution: Let X$\sim$ Laplace(a, b). Using the theorem in Chapter 1 (from Part I), we get

$$\text{MD} = 2\int_{ll}^{a} \text{F}(x)dx = 2\int_{-\infty}^{a} \frac{1}{2}\exp(-(a-x)/b)dx, \qquad (13.11)$$

where F(x) is the CDF. Put $z = (a - x)/b$, so that $dx = -b\,dz$. When $x = a$, $z = 0$, but when $x = -\infty$, z becomes $+\infty$. Cancel out the 2 to get

$$\text{MD} = -b\int_{\infty}^{0} e^{-z}dz = b\int_{0}^{\infty} e^{-z}dz = -be^{-z}\;\rvert_0^\infty = -b[0 - 1] = b. \qquad (13.12)$$

This shows that the mean deviation is b. The ratio MD/SD $= b/(\sqrt{2}b) = 1/\sqrt{2} = 0.70710678$.

Problem 13.4   Check whether $f(t) = C \exp(-1/(1 - (t/a)^2))$ for $-a < t < a$ is a PDF for a suitable choice of C.

13.4   FITTING

This distribution is symmetric around the location parameter, for which the sample median is the MLE. When the sample size n is even, the arithmetic mean of the middle values is taken as the estimate. Although the first derivative does not exist at a, the Cramer–Rao lower bound for a exists and is $1/[2(2n + 1)]$. Mean deviation from the median is the MLE of the shape parameter b. The variance of the median from a Laplace distribution is b^2/n, which approaches zero rapidly as $n \to \infty$ for fixed b, showing that it is also the MVUE. This can be used to form confidence intervals (CI) for the unknown location parameter.

13.5   EXTENSIONS OF LAPLACE DISTRIBUTION

Kozubowski and Nadarajah (2006) [121] identified over 16 variants of the Laplace distribution and summarized their statistical properties. Consider the distribution

$$f_x(a,b,c,r) = c \exp(-b|x - a|^r), \quad -\infty < x < \infty, \quad -\infty < a < \infty, \quad b > 0, \qquad (13.13)$$

where c is a normalizing constant. This reduces to the Laplace(a, b) for $r = 1$ and the normal law for $r = 2$. Intermediate values of r results in densities with tail decay in-between Gaussian and CLD distributions. Values of $r < 1$ results in even heavier tails. A more general form with PDF (Bottazzi et al. (2002) [25]) (who used $b' = 2/(1 + b)$)

$$\begin{aligned} f_x(a,b,c) = & 1/[c\Gamma(K)2^K] \exp(-\frac{1}{2}|(x - a)/c|^{2/(1+b)}), \\ & -\infty < x < \infty, \quad -\infty < a < \infty, \quad b > 0, \end{aligned} \qquad (13.14)$$

where $K = 1 + (1 + b)/2$, which reduces to the normal distribution for $b = 0$ and the Laplace distribution for $b = 1$. A variant of this with PDF

$$f_x(a,b,c) = c/[2b\Gamma(1/c)] \exp(-|(x - a)/b|^c), \quad -\infty < x < \infty, \quad -\infty < a < \infty, \quad b > 0 \qquad (13.15)$$

is used to model the photovoltaic power fluctuations using first-order differencing (Luo et al. (2018) [140]). As $E(X) = a$, a size-biased Laplace distribution can be obtained as

$$f(x; a, b) = (1/2ab) x \exp(-|x - a|/b), \quad -\infty < x < \infty, \quad -\infty < a < \infty, \quad b > 0, \qquad (13.16)$$

(alternatively, find the expected value of $(1 + cX)$ as in Chapter 1 (Part I)). A transmuted Laplace Distribution (TLD) can be obtained with the CDF

$$F(x; a, b) = ((1 + \lambda)G(x) - \lambda[G(x)]^2, \quad -\infty < x < \infty, \quad -\infty < a < \infty, \quad b > 0, |\lambda| \leq 1, \qquad (13.17)$$

where $G(x)$ is the CDF of one of the Laplace distributions.

13.5.1   ASYMMETRIC-LAPLACE DISTRIBUTIONS

The classical Laplace distribution is symmetric around the location parameter (a). An asymmetric Laplace distribution (ALD) or skew-Laplace distribution (SkLD) can be obtained by using different parameters for the tails as follows:

$$f(x;a,b,c) = \frac{ab}{(a+b)} \begin{cases} \exp(b(x-c)) & \text{for } x < c; \\ \exp(-a(x-c)) & \text{for } x \geq c, \end{cases}$$

where a and b describe the shape of left and right tail, $a > b$ results in thinner left tail and vice versa. This is also called two-piece Laplace distribution or skew double exponential distribution. This reduces to the SLD when $a = b$. The ALD gives better fit for empirical financial data than the Gaussian model. The two parts can be combined into a single expression using the indicator function as (Jones (2002) [106])

$$f(x;a,b,c) = \frac{ab}{(a+b)} \quad \exp(b(x-c))\mathrm{I}(x < c) - \exp(-a(x-c))\mathrm{I}(x \geq c). \tag{13.18}$$

As the left and right half tail off slowly (like the Pareto distribution), it is sometimes called double Pareto law. This property is critical for realistic modeling in several fields including geology, economics, finance and stock markets where the truncated asymmetric Laplace distribution (TALD) is better suited. This distribution is related to the log-Laplace distribution (discussed below) using $Y = \exp(X)$ where $X \sim$ALD(a,b,c) with $d = \exp(c)$. Casanova and Agnana (2000) [32] represented it using a single constant as

$$f(x;a,c) = a(1-a) \quad \exp(-(1-a)(c-x))\mathrm{I}(x < c) + \exp(-a(x-c))\mathrm{I}(x \geq c), \tag{13.19}$$

where $a \in (0,1)$. Holla and Bhattacharya (1968) [93] introduced a variant with PDF

$$f(x;a,b,p) = \begin{cases} p \quad b\exp(-b(a-x)) & \text{if } x < a; \\ (1-p) \quad b\exp(-b(x-a)) & \text{otherwise.} \end{cases}$$

This reduces to the SLD for $p = 1/2$.

Problem 13.5   Find the mean and variance of the above distribution.

Another parametrization as given below is also popular in some application areas:

$$f(x;a,b,c) = a(b+1/b)^{-1} \quad \exp(-(a/b)(x-c))\mathrm{I}(x < c) - \exp(-ab(x-c))\mathrm{I}(x \geq c). \tag{13.20}$$

Discrete version of ALD has been obtained by Kozubowski and Inusah (2006) [120] with PMF

$$f(x;p,q,b) = \frac{(1-p)(1-q)}{(1-pq)} \times \begin{cases} p^k & \text{for } k = 0,1,2,\ldots; \\ q^{|k|} & \text{for } k = -1,-2,\ldots, \\ \qquad p = \exp(-k/b), q = \exp(-1/(bk)) \end{cases} \tag{13.21}$$

which arises as the difference of two independent, but not identical geometric random variables, and has applications in foreign exchange rates for any two fixed currencies. This has ChF $\phi_x(t) = (1-p)(1-q)/[(1-p\exp(it))(1-q\exp(it))]$. The geometric distribution is related to discrete Laplace distribution analogously as the exponential distribution is related to Laplace distribution.

Problem 13.6 If $X \sim GEO(q)$ and $Y \sim GEO(q)$ is independent of X, prove that $U = X - Y$ is discrete Laplace distributed.

Problem 13.7 Find the mean and variance of a discrete Laplace distribution with PMF $f(x; p, c) = (q/(1+p)) \, p^{|x-c|}$, where $q = 1 - p$ and $x = c, c+1, \ldots$.

13.5.2 LOG-LAPLACE DISTRIBUTIONS

Consider the Pareto distribution described in the previous chapter and its reciprocal, which has a power function distribution. Log-Laplace Distribution (LLD) is defined by combining these two laws as follows:

$$f(x; a, b, d) = \frac{ab}{d(a+b)} \begin{cases} (x/d)^{b-1} & \text{for } 0 < x < d; \\ (d/x)^{a+1} & \text{for } x \geq d. \end{cases}$$

This has mean $abd/[(a-1)(b+1)]$, median $d(2K)^{-1/b}$ for $a \geq b$, and $d(2(1-K))^{1/a}$ for $a < b$ where $K = a/(a+b)$. This has CDF

$$F(x; a, b, d) = \begin{cases} K(x/d)^b & \text{for } 0 < x < d; \\ 1 - (1-K)(d/x)^a & \text{for } x \geq d, \end{cases}$$

where $K = a/(a+b)$. From this it is easy to find the quantiles as

$$F^{-1}(p; a, b, d) = d \begin{cases} (p/K)^{1/b} & \text{for } p \in (0, K]; \\ (q/(1-K))^{-1/a} & \text{otherwise.} \end{cases}$$

If $X \sim$ LLD (a, b, d), then $cX \sim$ LLD (a, b, cd) for $c > 0$. Another way to look at the LLD is as follows. Suppose $X \sim$ Laplace $(0, b)$. Then the distribution of $Y = \exp(X)$ has an LLD distribution. As $1/Y = \exp(-X)$, the reciprocal also is LLD. The product of a beta $(1, b)$ and an independent Pareto random variable is LLD. The hazard function (Chapter 1 of Part I) of LLD also is of the Pareto type, which is easy to prove using the relationship $h(x) = f(x)/[1 - F(x)] = f(x)/S(x) = -\frac{d}{dx} \log(S(x))$. A location parameter can be introduced to get a three-parameter version. This distribution and its reciprocal have the same distribution when $a = b$.

13.6 APPLICATIONS

This distribution finds applications in economics and finance, geology, environmental sciences, and various particle size modeling fields where it provides better fit than log-hyperbolic distribution. It is also used in modeling unreported data in demography, econometrics, and other fields (Hartley and Revankar (1974) [89]). It also has been extensively applied in growth-rate modeling (GDP, foreign exchange rates, etc.). If X_t denotes the size of an entity (like firm sizes, number of pages on WWW) at time t, and $Y_t = \log(X_t) - \log(X_{t-1})$ is the log-growth rate at time t, the distribution of Y_t can be approximated by the Laplace law. It finds applications in those fields where the rate of decay in both tails are slower than that of Gaussian and Student's t distributions. As an example, the biochemical oxygen demand exertion patterns in farm dairy waste-water of compost engineering can be approximated by a CLD (Mason, McLachlan, Gerard (2006) [146]).

It is also used to approximate the energy loss ratio as a function of measurement frequency under uncertainty in materials science, grain-size distribution during recrystallization in crystallography, wind shear modeling and wind power forecasting, and distribution of change rate of the stochastic component of historical capacity utilization in energy storage systems (Keles (2013) pages 119–120, [114]). Talha and Pätzold (2007) [212] used SLD to approximate the distribution of the envelope of double-scattering of fading channels in the absence of line-of-sight (LOS) component. Speech amplitude distribution is approximated using the Laplace law $f(x; a) = (a/2) \exp(-a|x|)$ in speech processing applications. The following paragraphs list some of the more popular applications.

13.6.1   ASTRONOMY

Consider observations of a celestial object (like a comet) using an array of instruments scattered throughout the world. Error heterogeneity is inevitable in such a situation due to the different characteristics of each observer, diversity of instruments used, differences in observational angles, and noise induced by extraneous variables in addition to relativistic differences at the point of site. A Laplace distribution gives a better model if each observer-instrument has its own variability.

13.6.2   NAVIGATIONAL ERRORS

A navigation system is used by an entity (like a robot, automobile, aircraft, or vessels of various sorts) to move to a certain destination on or above a terrain in a navigational area. Earth-based navigation systems quite often use one or more satellites in low-Earth orbits that send continuous radio signals which are captured by devices (receivers) within the moving entity to update its position. Cesium based highly accurate atomic clocks are used by modern satellites to get precise locational information in the receivers. Multiple satellites are needed for this purpose to cover the entire surface of the planet. However, positional errors (difference between actual and calculated position coordinates) are observed in large navigation systems due to a multitude of reasons like atmospheric interferences, multipath due to radio signals bouncing from both man-made and natural nearby objects like roof-tops, vehicles, slopes and walls, hills and mountains, tree-tops, etc., so that the received signal is a superimposition of multiple reflections of the same signal, received with varying delays and attenuation.

A time-varying multipath that results from the motion of the transmitter, receiver, and some of the reflectors (like vehicles) adds to the woes due to a Doppler shift in the received signal frequency. Even a few nanosecond delays can result in large positional errors. At least three satellites should be in the visible horizon to get accurate coordinate information in global positioning systems (GPS). Deterioration in signal quality is a serious problem in under-water missions, mines, and basement floors of large buildings (where some parking lots are located). Getting highly accurate position coordinates is crucial in some navigation systems used by aircraft and ships to avoid disastrous collisions, in battlefields and missile hits, in search and rescue missions, and in underwater repair robots. Commercial aircraft usually use parallel tracks that increase the mid-air collision chances. Due to the surge in unmanned aerial vehicles (UAV) and drone delivery systems in metro areas, it is gaining popularity in collision avoidance of such systems too. McFadyen and Martin (2018) [150] used a non-zero mean CLD to model the altitude keeping ability. They also used a convex combination of two independent CLD components having the same mean but different shape parameters with respective weights α and $(1-\alpha)$ to model typical and atypical behaviors.

Positioning error distribution is a well-studied problem in every navigation system (Hsu (1979) [95]). In fact, the systems are calibrated during the inception (testing phase) using position errors calculated from empirical data of known coordinates. Accuracy of GPS-based systems will drastically decrease when one or more satellites are eclipsed due to high rise buildings or other obstacles in the receivers way. This usually happens in road-vehicles moving in large cities. Position errors can be envisaged either in the calculated coordinates or separately in the individual dimensions of 3D space. In the later case, errors are assumed to be unimodal with mode at zero. One example is the lateral errors of aircraft in motion along parallel tracks.

Student's t and Laplace distributions are the preferred choice in modeling position errors due to simplicity and flexibility. Anderson and Ellis (1971) [4] observed that the tail region of the frequency distribution of position errors accumulated from multiple sources resemble that of an exponential function of the magnitude of the error with PDF $f(x; a, b) = (1/2b)\exp(-|x|/b)$. In an attempt to establish a relation between Laplace and Gaussian laws, Hsu (1979) [95] assumed that the position errors are Gaussian $N(0, \sigma^2)$ distributed with an unknown prior distribution for σ^2, and thereby showed the prior to be the exponential law $EXP(2b^2)$ from which he obtained the compound distribution as the zero-mean Laplace law.

13.7   SUMMARY

This chapter introduced the Laplace distribution and its extensions. Whenever empirical data appears like an exponential distribution "spliced together back-to-back," it is an indication that the Laplace distribution is a better choice. Special Laplace distributions like truncated, asymmetric, and log-Laplace distributions are briefly introduced. The chapter ends with a list of applications of Laplace distribution in engineering and scientific fields.

CHAPTER 14

Central Chi-squared Distribution

14.1 INTRODUCTION

This distribution has a long history dating back to 1838 when Bienayme obtained it as the limiting form of multinomial distribution. It plays an important role in testing hypotheses about frequencies and counts captured in contingency tables, as shown in the applications section (page 209). It was used by Karl Pearson for contingency table analysis during 1900. It is also used in testing goodness of fit between observed data and predicted model and in constructing confidence intervals for sample variance.

If $X_1, X_2, \ldots, X_n$ are independent N(0, 1) random variables, the distribution of $Y = X_1^2 + \cdots + X_n^2$ is called the central χ^2 distribution. It has only one parameter called degrees of freedom (n) and has PDF

$$f(x; n) = x^{n/2-1} e^{-x/2} / [2^{n/2} \Gamma(n/2)], \quad x, n > 0. \tag{14.1}$$

It can also be written as

$$f(x; n) = (1/[2\Gamma(n/2)])(x/2)^{n/2-1} e^{-x/2}, \quad x, n > 0. \tag{14.2}$$

It is a special case of the Gamma distribution GAMMA($n/2, 1/2$). An alternate notation used in some applied science fields is

$$f(\chi^2; n) = (\chi^2)^{n/2-1} e^{-\chi^2/2} / [2^{n/2} \Gamma(n/2)], \quad n > 0. \tag{14.3}$$

We denote it by χ_n^2, $\chi^2(n)$ or CHISQ(n), where the DoF (which is usually lowercase English letters like n, m or Greek letter ν) is written either as a subscript or as argument.[1] The test statistic that uses this distribution is denoted simply by χ^2, and aptly called chi-square statistics. It is also called *central chi-squared* distribution. The "square" part in the etymology indicates the fact that it is related to squares of IID normal variates with zero means. It finds extensive applications in testing of hypotheses and confidence intervals in the frequentist approach (tests for variance, goodness of fit, independence, consistency, etc.), contingency tables, quality control, reliability, communication engineering, etc. It is used in cryptanalysis to compare significance difference, if any, between plaintext and decrypted ciphertext.

---

[1]Some authors denote the DoF by ν because n usually indicates the sample size in statistical inference. This distribution is sometimes called Helmert's [91] distribution.

This distribution can be derived geometrically using the probability content of regions under spherical normal distributions (Ruben (1964) [188]).

14.2   RELATION TO OTHER DISTRIBUTIONS

It is the distribution of sum of squares of IID standard normal random variables. Symbolically, if $Z_1, Z_2, \ldots, Z_n$ are IID N(0, 1) variates, then $\sum_{k=1}^{n} Z_k^2 \sim \chi_n^2$. A direct extension of this result to the multivariate case is as follows. If $X \sim N_p(0, I)$ is p-variate normal, then the quadratic form $Q = X'X$ can be decomposed into k quadratic forms $Q_j = X'B_j X$, for $j = 1, 2, \ldots, k$ where the rank of B_j is r_j, and each B_j is positive semi-definite, then each $Q_j \sim \chi_{r_j}^2$, and all Q_j's are mutually independent (this is called Cochran's theorem in statistics). As a particular case, if $X \sim N(0, 1)$, then $X^2 \sim \chi^2(1)$ distribution. Similarly, if $X \sim N(\mu, \sigma^2)$, then $[(X - \mu)/\sigma]^2$ is $\chi^2(1)$ distributed. A scaled χ^2 distribution has PDF

$$f(x; n, c) = (1/[2c^2\Gamma(n/2)])(x/(2c^2))^{n/2-1}e^{-x/(2c^2)}, \quad n > 0, \qquad (14.4)$$

which is Gamma $(x/(2c^2), n/2)$. It is also related to Stacy distribution as Stacy $(x/(2c^2), k/2, 1)$. The sampling distribution of sample variance is χ^2 distributed when data are IID normal. If $X \sim U(0, 1)$ then $Y = -2\log(X) \sim \chi_2^2$ distributed. The distribution of positive square root of a χ^2 random variable is known as Chi distribution and has PDF

$$f_x(n) = x^{n-1}e^{-x^2/2}/[2^{n/2}\Gamma(n/2)]. \qquad (14.5)$$

Rayleigh distribution, half-normal distribution, and Maxwell distributions are related to the χ distribution. The distribution of sample standard deviation from a normal population has a scaled χ-distribution. The ratio of two IID χ^2 variates divided by their respective DoF has an F distribution. Symbolically, $Y = (\chi_m^2/m)/(\chi_n^2/n) \sim F(m, n)$, if the numerator and denominator are independent. Similarly, $X/(X + Y) \sim$ Beta-I$(m/2, n/2)$, $Y/(X + Y) \sim$ Beta-I$(n/2, m/2)$, and $X/Y \sim$ Beta-II$(m/2, n/2)$, where X and Y are IID χ^2-variates. The Nakagami-m distribution used in communication engineering is a scaled χ distribution with PDF

$$f(x; m, \Omega) = 2m^m x^{2m-1}/[\Omega^m \Gamma(m)]\exp(-mx^2/\Omega). \qquad (14.6)$$

If $X \sim$ IGD(μ, λ), then $Y = \lambda(X - \mu)^2/(\mu^2 X)$ has χ_1^2 distribution. If $X \sim F(m, n)$ is the central F distribution, then $Y = \lim_{n\to\infty} mX$ is central χ_m^2 distributed. If $X \sim \chi_n^2$ and $c > 0$ is a constant, then $Y = cX$ has a GAMMA $(n/2, 2c)$ distribution. The distribution of χ^2 with $2n$ DoF is a special Erlang distribution with parameters $(n, 1/2)$. Alternatively, if $X \sim$ Erlang(n, λ), then $Y = 2\lambda X \sim \chi_{2n}^2$. If $X_k \sim$ Laplace(a, b) for $k = 1, 2, \ldots, n$, then $Y = (2/b)\sum_{k=1}^{n} |x_k - a| \sim \chi_{2n}^2$. It is also related to the inverse Gaussian distribution IG(μ, λ) as $\lambda \sum_{k=1}^{n}(1/x_k - 1/\bar{x}) \sim \chi_{n-1}^2$ distributed. If $X_1, X_2, \ldots, X_n$ are IID N(μ_k, σ^2) random variables, the distribution of $Y = \sum_k (X_k - \bar{x})^2/\sigma^2$ has a χ^2 distribution with $n - 1$ DoF. This could be expressed in terms of the sample variance as $(n - 1)s^2/\sigma^2$, which forms the basis for tests of hypotheses on population variance (in single-sample and multi-sample cases). The noncentral χ^2 distribution is an extension in which at least one of the Gaussian components has non-zero mean. It reduces to the central χ^2 when the noncentrality parameter is zero.

Expressions of the form $Q = (X - \mu)'A(X - \mu)$ where X is a k-dimensional Gaussian vector and A is a symmetric non-negative matrix appear in various applied fields. For example, sequential χ^2 criterion for multivariate comparison of means from a multivariate normal population using a probability ratio requires the value of $\chi_n^2 = n(\overline{X}_n - \mu_0)'\Sigma^{-1}(\overline{X}_n - \mu_0)$ where $\overline{X}_n$ is the sample mean vector, and population covariance matrix Σ is assumed to be known. The PDF of multivariate normal can be written as

$$\phi(x; \mu, \Sigma) = (1/(2\pi))^{p/2}/|\Sigma|^{1/2} \quad \exp(-\frac{1}{2}((x - \mu)'\Sigma^{-1}(x - \mu))), \qquad (14.7)$$

where p is the number of variates (dimensionality). This PDF depends on x through the exponent $(x - \mu)'\Sigma^{-1}(x - \mu)$, which is called the population squared Mahalanobis' distance. Fixing this quantity as $(x - \mu)'\Sigma^{-1}(x - \mu) = c$ results in a hyper-ellipsoid in p-dimensional space.

Theorem 14.1   *If $X \sim N(0, I_p)$ is a standard multivariate normal variate, then $Q = X'X \sim \chi_p^2$ distribution.*

Proof. If X is a k-dimensional Gaussian vector variate $N(\mu, \Sigma)$, then the population Mahalanobis distance (PMD) $\Delta^2 = (X - \mu)'\Sigma^{-1}(X - \mu)$ is distributed as χ_k^2, where k is the rank of the variance-covariance matrix Σ. As Σ is a symmetric matrix, the Δ^2 is preserved under full-rank linear transformation of the variates. An eigenvalue decomposition of Σ^{-1} can be used to obtain the distribution as follows. Let $\Sigma^{-1} = U\Lambda^{-1}U'$ be the eigen-decomposition where U is an orthonormal matrix, and Λ is diagonal. Thus, we could express $\Delta^2 = (X - \mu)'\Sigma^{-1}(X - \mu)$ as $\Delta^2 = \sum_{k=1}^{p} \lambda_k^{-1}(X - \mu)'u_k u_k'(X - \mu) = \sum_{k=1}^{p} \lambda_k^{-1}\left(u_k'(X - \mu)\right)^2$. Combine λ_k^{-1} with the square to get $\Delta^2 = \sum_{k=1}^{p} \left(\lambda_k^{-1/2}u_k'(X - \mu)\right)^2$. Now use the transformation $Y_k = \lambda_k^{-1/2}u_k'(X - \mu)$ to get $\Delta^2 = \sum_{k=1}^{p} Y_k^2$. As linear transformations of Gaussian variates are Gaussian distributed, each of the $Y_k \sim N(0, \sigma_k^2)$, and the joint distribution is $N(0, \Sigma)$, so that Δ^2 is the sum of independent χ^2 random variates. Using the additiviy property, it follows that $\Delta^2 \sim \chi_p^2$ distributed.

The sample measure $T^2 = N(\overline{X} - \mu)'S^{-1}(\overline{X} - \mu)$ is called Hotelling's T^2 statistic. Under normality assumption, $(N - n)/[n(N - 1)]T^2$ is distributed as $F(n, N - n)$, for $N > n$ where N is the sample size and n is the dimensionality (number of variables). A necessary and sufficient condition for $X'AX$ to be χ^2 distributed is that the matrix A is idempotent (i.e., $A^2 = A$) in which case the DoF of χ^2 is the trace of A. Finding tail areas of such quadratic forms is not easy except in special cases (e.g., when A is a diagonal matrix). Kotz et al. (1967) [119] expressed it as an infinite series of central χ^2 random variables, whereas Liu, Tang, and Zhang (2009) [139] used a weighted sum of χ^2 random variables, for which easy approximations are available (Gabler and Wolff (1987) [75]). Mardia's test for multivariate skewness uses the test statistic $b_{1,p} = (1/n^2)\sum_{j=1}^{n}\sum_{k=1}^{n}[(x_j - \overline{x})'S^{-1}(x_k - \overline{x})]^3$, which for $p > 2$ can be approximated using a central χ^2 distribution (Mardia (1974) [144]).

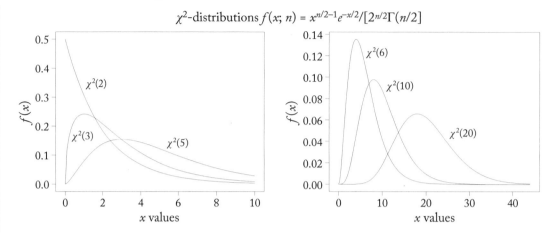

Figure 14.1: Chi-squared distributions.

14.3   PROPERTIES OF χ^2 DISTRIBUTION

Although the DoF parameter n is an integer in many applications, it can be any real number theoretically. As $n \to \infty$, this distribution approaches normality (Figure 14.1). Better convergence rates are observed for some nonlinear transformations. For example, $\sqrt{2X}$ is approximately normal with mean $\sqrt{2n-1}$ and variance 1 (Fisher (1931) [67]), whereas the cube-root transformation $(X/n)^{1/3}$ is approximately normal with mean $(1 - 2/(9n))$ and variance $2/(9n)$ (Wilson and Hilferty (1931) [229]). Chattamvelli and Shanmugam (1998) [44] obtained a normal approximation to the CDF of noncentral beta distribution using this result.

Integrate (14.1) wrt x to get the CDF as

$$F_n(x) = (1/[2^{n/2}\Gamma(n/2)]) \int_0^x y^{n/2-1} e^{-y/2} dy. \tag{14.8}$$

Change the variable as $x = y/2$ to get (14.8) in terms of incomplete gamma function as

$$F_n(x) = \gamma(n/2, x/2)/\Gamma(n/2) = P(n/2, x/2), \tag{14.9}$$

where $P(u, v)$ is the regularized gamma function. The corresponding relationship for χ-distribution is

$$F_n(x) = \gamma(n/2, x^2/2)/\Gamma(n/2) = P(n/2, x^2/2), \tag{14.10}$$

The CDF of χ_n^2 satisfies the recurrence

$$F_n(x) - F_{n-2}(x) = -2 * f_n(x). \tag{14.11}$$

Alam and Rizvi (1967) [2] extended (14.11) to the noncentral case. Chattamvelli (1995b) [37] represented it as $(1 + \Delta)F_n(x) = F_{n-2}(x)$, where $\Delta = 2\partial/(\partial x)$, and obtained an elegant derivation of Han's (1978, 1979) [86, 87] result using a similar differential involving the non-centrality parameter. Chattamvelli and Shanmugam (1995) [43] obtained sharp error bounds for truncating infinite series expansions for the CDF of noncentral χ^2 distributions using (14.11). Repeated application of (14.11) gives a simple expression for the CDF with $n + 2k$ DoF in terms of the CDF for n DoF as

$$F_{n+2k}(x) = F_n(x) - f_n(x) \sum_{j=1}^{k} (x/2)^j / (n/2)_j, \tag{14.12}$$

which is useful in computing percentage points for large DoF (Fisher (1931) [67]). For instance, the CDF of χ_n^2 for odd DoF can be found using CDF of χ_1^2, which has a representation in terms of standard normal CDF as $F_1(x) = \Phi(\sqrt{x}) - \Phi(-\sqrt{x})$ (Han (1978) [86] and Chattamvelli (1995) [37] extended this to the noncentral case), and for even DoF in terms of bivariate normal CDF as $F_2(y) = (2/\pi) \int_0^{\sqrt{y}} \int_0^{\sqrt{y-x^2}} \exp(-\frac{1}{2}(x^2 + z^2)) dz\, dx$ (Han (1979) [87]). Bock and Govindarajulu (1989) [23] represented it in terms of modified Bessel functions.

Substituting for the PDF results in a second-order recurrence for the CDF as

$$(n - 2)F_n(x) = (x + n - 2)F_{n-2}(x) - x F_{n-4}(x), \tag{14.13}$$

(Ruben (1964) [188], Chattamvelli and Jones (1995) [42]). Ruben (1964) and Cameron (1974) [31] expressed (14.12) as

$$F_{n+2k}(x) = F_n(x) - 2f_n(x) \sum_{j=1}^{k} f_{n+2j}(x). \tag{14.14}$$

Johnson (1959) [103] obtained a relation between the CDF of central χ^2 distribution and SF of Poisson distribution as $\Pr[\chi_n^2 \leq x] = \Pr[Y \geq n/2]$ where $Y \sim POIS(x/2)$, which he extended to the noncentral case. Jones (1987) [105] applied the Poisson-exponential model to insurance claim data in which claims arrive according to a Poisson process, and the size of the claims follow IID exponential random variables.

14.3.1   MOMENTS AND GENERATING FUNCTIONS

The MGF is easily obtained as

$$M_x(t) = E(e^{tx}) = \int_0^{\infty} e^{tx} x^{n/2-1} e^{-x/2} / [2^{n/2} \Gamma(n/2)] dx \tag{14.15}$$

$$= 1/[2^{n/2} \Gamma(n/2)] \int_0^{\infty} e^{-x[1-2t]/2} x^{n/2-1} dx = (1 - 2t)^{-n/2}.$$

From this the kth moment is easily obtained as $\mu_k' = 2^k \Gamma(n/2 + k)/\Gamma(n/2)$. The mean is n and variance is $2n$. The χ^2 family of distributions is over-dispersed as the variance is twice the mean. The mode is $n - 2$ for $n > 2$, and zero otherwise. The cumulants are $\kappa_n = 2^{k-1}(k - 1)!n$, which satisfies

Table 14.1: Properties of χ^2 distribution $x^{n/2-1}e^{-x/2}/[2^{n/2}\Gamma(n/2)]$

Property	Expression	Comments
Range of X	$0 \leq x < \infty$	Continuous
Mean	$\mu = n$	
Mode	$n - 2 = \mu - 2$	$n > 2$
Median	$n - (2/3) = \mu - 2/3$	$\approx n(1 - 2/(9n))^3$
Variance	$\sigma^2 = 2n = 2\mu$	$\sigma^2 > \mu$
CV	$(2/n)^{1/2}$	
Skewness	$\gamma_1 = 2\sqrt{2/n}$	$= 2^{3/2}n^{-1/2}$
Kurtosis	$\beta_2 = 3 + 12/n$	Always leptokurtic
Mean deviation	$e^{-\frac{n}{2}}n^{\frac{n}{2}+1}/\left[2^{\frac{n}{2}-1}\Gamma\left(\frac{n}{2}+1\right)\right]$	$= 2\int_0^n P(n/2, x/2)dx$
Moments	$\mu_r' = 2^r\Gamma(r + n/2)/\Gamma(n/2)$	
MGF	$(1 - 2t)^{-n/2}$	$t < 1/2$
ChF	$(1 - 2it)^{-n/2}$	
CGF	$-(n/2)\log(1 - 2it)$	
Additivity	$\chi_m^2 + \chi_n^2 = \chi_{m+n}^2$	Independent
Recurrence	$f_{n+2}(x)/f_n(x) = x/n$	$F_n(x) - F_{n-2}(x) = -2f_n(x)$
Approx	$(\chi_n^2/n)^{1/3}$	$N((1 - 2/(9n)), 2/(9n))$
Tail probability	$1 - P(n/2, x/2)$	P = Regularized gamma f_n

a recurrence $\kappa_{n+1} = 2k\kappa_n$. See Table 14.1 for further properties. The moments of χ-distribution is

$$E(x^k) = E[(x^2)^{k/2}] = 2^{k/2}\Gamma((n + k)/2)/\Gamma(n/2). \tag{14.16}$$

14.3.2   MOMENT RECURRENCES

Moment recurrences are most easily obtained from the corresponding density recurrences. The basic density recursion is $nf_{n+2}(x) = xf_n(x)$, which shows that the χ_{n+2}^2 distribution is obtained by size-biasing a χ_n^2 distribution. A recurrence relation for the raw moments of the central χ^2 is

$$n\mu_{n+2}^k = \mu_n^{k+1}, \tag{14.17}$$

which in the noncentral case $(\chi_n^2(\lambda))$ is $n\mu_{n+2}^k(\lambda) = \mu_n^{k+1}(\lambda) - \lambda\mu_{n+4}^k(\lambda)$ (Chattamvelli and Jones (1995) [42]).

14.3.3   ADDITIVITY PROPERTY

This distribution in particular, and its noncentral version in general satisfies a reproductive property. If $X \sim \chi_m^2$ and $Y \sim \chi_n^2$ are independent, then $X + Y \sim \chi_{m+n}^2$. This result was proved by Helmert (1876) [91] for the central χ^2 distribution, but could be extended to the noncentral case using the MGF or ChF. If $X_1, \ldots, X_k$ are independent random variables, X_i being $\chi^2(n_i)$ distributed, then $X_1 + \cdots + X_k$ is a $\chi^2(n_1 + \cdots + n_k)$ random variable. This result is used to find the distribution of quadratic forms (Mathai and Provost (1992) [147]), and in least squares theory (Rao (1973), p. 189; [180]). Let $X = (X_1, X_2, \ldots, X_n)$ be a vector of IID N(0, 1) variates, and $Y = x'Ax$ be a quadratic form where A is a symmetric matrix. If A is of full-rank, an orthogonal transformation $x = Ty$ can be used to reduce the symmetric matrix to diagonal form where the diagonal element are the eigen values of A. Then $x'Ax = y'T'ATy = \sum_{k=1}^{n} \lambda_k y_k^2$. As X_k's are IID unit normal variates, so are Y_k's (Chapter 8 of Part I). Thus, the distribution of quadratic form is that of a weighted linear combination of independent χ^2 variates, which has a gamma distribution. If X_1, X_2 are IID $\chi^2(n)$ distributed, then $Y = aX_1 + bX_2$ has a Bessel-function distribution for $a > b > 0$ with PDF

$$f(y) = K \, y^m e^{-cy/d} \, I_m(y/d), \quad \text{for } y > 0, \tag{14.18}$$

where $c = (a + b)/(a - b), d = 4ab/(a - b), m = 2n + 1, I_m(y)$ is the modified Bessel function of first kind, and the normalizing constant $K = (c^2 - 1)^{m+.5}/[\sqrt{\pi}2^m d^{m+1}\Gamma(m + .5)]$.

14.3.4   APPROXIMATIONS

Wilson and Hilferty (1931) [229] proved that $(\chi_n^2/n)^{1/3}$ is approximately normal with mean $(1 - 2/(9n))$ and variance $2/(9n)$. This allows the CDF to be expressed in terms of standard normal CDF as

$$F_n(x) = \Phi(((x/n)^{1/3} - 1 + 2/(9n))/\sqrt{2/(9n)}). \tag{14.19}$$

Many other approximations are available. See, for example, Narula and Franz (1977) [164], Zar (1978) [238], Lin (1988) [138], and Robert (1990) [185].

Problem 14.2   If $X \sim \chi_n$ (i.e., $X \sim \sqrt{\chi_n^2}$) and $Y \sim$ BETA $(\frac{n-1}{2}, \frac{n-1}{2})$ is independent of X, prove that $(2Y - 1)X \sim N(0, 1)$.

Problem 14.3   If $X \sim \chi_m^2, Y \sim \chi_n^2$ and $Z \sim \chi_p^2$ are independent, find the distribution of $U = X/Y, V = (X + Y)/Z$ and $W = X + Y + Z$.

Example 14.4   Mean deviation of Central chi-square distribution   Find the mean deviation of the central χ^2 distribution.

Solution: We apply power method ([47]) to find the MD. As the CDF is $P(n/2, x/2)$, and mean $\mu = n$, we get

$$\text{MD} = 2 \int_0^n P(n/2, x/2)dx. \tag{14.20}$$

Use integration by parts by taking $u = P(n/2, x/2)$, $dv = dx$ to get

$$MD = 2nP(n/2, n/2) - 2/[2^{n/2}\Gamma(n/2)] \int_0^n x \; x^{n/2-1} e^{-x/2} dx. \tag{14.21}$$

Put $x/2 = u$ so that $dx = 2du$ to get

$$MD = 2nP(n/2, n/2) - 4/\Gamma(n/2) \int_0^{n/2} u^{n/2} e^{-u} du. \tag{14.22}$$

Multiply the numerator and denominator of the second term by $(n/2)$ and write the denominator as $(n/2) * \Gamma(n/2) = \Gamma(n/2 + 1)$. This gives

$$MD = 2n[P(n/2, n/2) - P(n/2 + 1, n/2)]. \tag{14.23}$$

Now use $P(a, x) - P(a + 1, x) = e^{-x} x^a / \Gamma(a + 1)$ with $a = x = n/2$ to get

$$MD = 2ne^{-n/2}(n/2)^{n/2}/\Gamma(n/2 + 1). \tag{14.24}$$

This simplifies to

$$MD = e^{-n/2} n^{n/2+1}/[2^{n/2-1}\Gamma(n/2 + 1)]. \tag{14.25}$$

Write $\Gamma(n/2 + 1) = (n/2)\Gamma(n/2)$ and cancel out n to get

$$MD = e^{-n/2} n^{n/2}/[2^{n/2-2}\Gamma(n/2)]. \tag{14.26}$$

Problem 14.5 If F(x) is the CDF of a continuous random variable defined on $[0, 1]$, prove that $-2\ln(F(x))$ has a χ^2 distribution with 2 DoF.

14.4 FITTING

As the mean of the distribution is n, the MLE and MOM estimators of n are the sample mean. Thus, it is very easy to fit a χ^2 distribution. As the domain of this distribution is $[0, \infty]$, the data used for fitting must all be positive. However, the mean of such data can be any positive real number (not necessarily integer). As noted before, the PDF is defined even for non-integer values of n although it denotes the DoF.

14.5 EXTENSIONS OF χ^2 DISTRIBUTION

A direct extension of χ^2 distribution is the noncentral χ^2 distribution which is the distribution of the sum of squares of n IID normal random variables with at least one non-zero mean. A transmuted χ^2 distribution (TCD) can be obtained with the CDF

$$F(x; a, b) = (1 + \lambda)G(x) - \lambda[G(x)]^2, \quad -\infty < x < \infty, \quad -\infty < a < \infty, \quad b > 0, |\lambda| \le 1, \tag{14.27}$$

where $G(x)$ is the CDF of one of the χ^2 distributions.

14.5.1 SIZE-BIASED χ^2 DISTRIBUTIONS

As the mean of a χ^2 distribution is n, a size-biased χ^2 distribution can be obtained as

$$f(x;n) = x^{n/2}e^{-x/2}/[n\ 2^{n/2}\Gamma(n/2)] = (x^2)^n e^{-x/2}/[n\ 2^{n/2}\Gamma(n/2)], \qquad (14.28)$$

which belongs to the same family. Alternately, take a linear function $(1 + cx)$ with expected value $(1 + cn)$, where c is a constant. From this, a size-biased PDF follows as

$$f(x;c,n) = (1 + cx)x^{n/2-1}e^{-x/2}/[(1 + cn)\ 2^{n/2}\Gamma(n/2)]. \qquad (14.29)$$

14.6 APPLICATIONS

This distribution finds extensive applications in statistical inference, especially in tests of hypotheses and confidence intervals. As an example, several research studies in epidemiology involve comparing the significance of nominal categories. One situation is to check whether there is any significant difference for coronavirus mortality among males, females, and transgenders in various categories like different ethnic groups, countries, etc. It can be used to test independence of two qualitative attributes in contingency tables, goodness of fit of observed and hypothesized distributions, in likelihood-ratio tests (LRT), Neyman-Pearson framework, minimum χ^2-method (MCM) in model selection, χ^2-plots to check joint normality of data, etc. A premise of the χ^2-test is that the deviations observed from what is to be expected are purely due to statistical fluctuations. As the χ^2 distribution is additive, the results from many independent data sets may be combined and tested together.

The large-sample distribution of generalized LRT is approximately χ^2 distributed. As the normal and chi-square approximations for LRT are valid for large sample sizes only, it is preferable to use the t distribution for $n < 30$, and χ^2 otherwise. χ^2 tests are also used in Bartlett's approximation for Wilk's λ, Bartlett's test for equality of k variances, sphericity test, etc. Correlated χ^2 random variables are encountered in communication engineering, robotic motion planning, and many other fields. See Gordon and Ramig (1983) [82] for the CDF of sums of correlated χ^2 variates. The SNR is χ^2 distributed when Rayleigh model is used as the flat fading model in a receiver that receives signal energy equally through N sub-paths that fade independently.

SQC applications use $C_p = (USL - LSL)/(6s)$ where USL and LSL are the upper and lower control limits, and s is the sample standard deviation. Confidence intervals for C_p are based on the χ^2 distribution as $\hat{c}_p \chi^2_{n-1}(1 - \alpha/2)/(n - 1) \leq C_p \leq \hat{c}_p \chi^2_{n-1}(\alpha/2)/(n - 1)$ where n is the sample size, and $\hat{c}_p$ is the estimated value of C_p. The quantity SSE/σ^2 encountered in multiple regression and other least squares problems is approximately χ^2 distributed. It is assumed that the test statistic has a χ^2 distribution under the null hypothesis H_0 in all of the tests discussed below.

14.6.1 χ^2-TEST FOR VARIANCE

Suppose a random sample of size n is drawn from a population with unknown variance. To test the hypothesis $H_0 : \sigma^2 = \sigma_0^2$ against the alternative when the population mean is known (say μ), we use the statistic $\chi^2 = \sum_{k=1}^{n}(x_k - \mu)^2/\sigma_0^2$ which has a χ^2 distribution with n DoF when the population is normal. If the population mean is unknown, the sample mean is substituted and the χ^2 statistic becomes $\sum_{k=1}^{n}(x_k - \bar{x})^2/\sigma_0^2 = (n-1)s^2/\sigma_0^2$, which has a χ^2 distribution with $(n-1)$ DoF. The critical region depends on whether the alternate hypothesis is one-sided or double-sided. Confidence intervals for unknown population variance uses the χ^2 tail areas as $[(n-1)s^2/b, (n-1)s^2/a]$ where $\Pr[a < \chi_{n-1}^2 < b] = 1 - \alpha$ can be obtained from χ^2 tables or computed numerically.

14.6.2   CONTINGENCY TABLES

A contingency table (also known as a bivariate table) is a rectangular array of numbers in which the entries are counts or frequencies (and not percentages, proportions, or fractions), arranged into rows and columns. Each entry is uniquely identified by its row and column numbers (indices, usually denoted by subscripts). In most applications, we assume that N objects are randomly chosen from a population and arranged according to the row and column categories so that the corresponding cells contain the frequencies. Such a cross-tabulation presents the distribution of sampled items into two or more mutually exclusive and independent categorical variables simultaneously. Note that each data item contributes to one cell only (meaning that data values that fall on the boundary are unambiguously assigned to a unique category when the classes are obtained by categorizing quantitative variables, as in decision tree induction of machine learning). Such an arrangement is called a contingency table (CT) in statistics and allied fields. Each CT is identified by the number of rows and columns, and denoted as $CT(m, n)$, $CT_{m \times n}$, or $CT(m \times n)$ where m is the number of rows, and n the number of columns, where $\min(m, n) \geq 2$. The order in which rows and columns appear is unimportant. What actually matters in the analysis is the numbers that appear in a CT. There exist only two categories for one of the variables in several research studies (like real medicine vs. placebo, infected vs. not-infected, etc.) that results in $CT(2 \times n)$.

It is an implicit assumption in some of the tests discussed below that each entry is at least five. This is because the χ^2-test statistic is obtained by squaring the normalized binomial distribution $(x - np)/\sqrt{npq}$, and the normal approximation to binomial distribution is valid only for large n (see below). Moreover, instability results as the expected frequencies (in the denominator) tends to zero. Adjacent rows or columns may be combined when this condition is violated, if doing so will increase each frequency to be higher than five.

Suppose a researcher is interested in finding out whether two samples have come from the same population or not. There are many statistical tests to check this. Perhaps the most well-known among them is Pearson's χ^2 test, the data for which comes from a CT. If the conditions mentioned above are satisfied, it is a simple matter to compute the test statistic described in the following paragraphs. Note that the observed counts (O_i) are all non-negative integers, whereas the expected counts (E_i) can be fractions also. A Yate's correction may be applied when at least one of the expected counts is non-integer by modifying the numerator as $(|O_i - E_i| - 0.5)^2$ instead of $(O_i - E_i)^2$.

14.6.3 χ^2-TEST FOR INDEPENDENCE

This test is used to decide whether two variables (factors) are independent (null hypothesis) or dependent (alternate hypothesis). Let X be a binomial random variable BINO(n,p) which has mean np and variance npq [46]. Then it is well known from the time of De Moivre and Laplace that $Z = (x - np)/\sqrt{npq} \to N(0,1)$ for large n. Squaring gives $Z^2 = (x - np)^2/npq \sim \chi_1^2$. As $1/(pq) = 1/p + 1/q = 1/p + 1/(1 - p)$, we have $(x - np)^2/(npq) = (x - np)^2/(np) + (x - np)^2/(n(1 - p))$. Write $(x - np)^2 = (np - x)^2 = ((n - x) - nq)^2$ in the second expression to get $Z^2 = (x - np)^2/(np) + ((n - x) - n(1 - p))^2/(n(1 - p))$. This has the implication that we are "squaring the expected value from observed count in each of the success and failure categories to get a test-statistic." This is the starting point of Pearson's χ^2 test and its variants. If the row totals are denoted by n_i, and column totals by c_j so that $\sum_i n_i = \sum_j c_j = N$, the expected values are given by $E_{ij} = n_i c_j/N$. The null hypothesis is $H_0 : p_{ij} = p_{i.}p_{.j}$, which signifies that the attributes are independent.

When N data values are arranged in the form of a table with r rows and c columns, this becomes $Z^2 = \sum_{i=1}^r \sum_{j=1}^c (O_{ij} - E_{ij})^2/E_{ij}$, which has a χ^2 distribution with $(r - 1)(c - 1)$ DoF. Expand the quadratic term, and simplify to get an alternate expression $T = \sum_{i=1}^r \sum_{j=1}^c (O_{ij}^2/E_{ij}) - N$. Chi-square test thus quantifies the observed deviations from that expected by chance under the null hypothesis. It is assumed that each of the $E_{ij}'s \geq 5$. The χ^2 statistic is computed from the cross-tabulated data and compared with a critical value from the chi-square tables. Conclusions are drawn based on whether the observed cell counts differ significantly from the expected cell counts using the magnitude of Z^2. Most of these tests are right-tailed, which means that the null hypothesis is rejected if the computed value is greater than the tabulated value.

Several extensions of this test exists. For example, we could test if probabilities differ when samples are drawn from r populations in which each item can be classified into one of c classes. Another χ^2 statistic defined as $\chi_\lambda^2 = 2 \sum o_i \ln(o_i/e_i)$ when minimized gives the same estimate as MLE. The χ^2-test for consistency is a special case that tests the significance of k different populations. In the case of dichotomous population, the test statistic is $\chi^2 = (n - 1)(ad - bc)^2/[(a + b)(a + c)(b + d)(c + d)]$ where a, b, c, d denotes the observed frequencies falling in the two classes for each population. This has a χ_1^2 distribution, so that the hypothesis is rejected if computed value is greater than tabulated value.

A 2×2 chi-square test for independence is called Fisher's exact test. When samples come from k different populations with a dichotomous classification (two classes), the test statistic becomes $\chi^2 = (N^2/[S(N - S)])(\sum_{j=1}^k x_j^2/n_j - S^2/N)$ where n_j are the total counts from sample j, x_j and $n_j - x_j$ denote the count of jth sample falling under the first and second classes, $X = \sum_j x_j$ and $N = \sum_j n_j$. This has a χ^2 distribution with $k - 1$ DoF.

14.6.4 χ^2-TEST FOR GOODNESS OF FIT

This nonparametric test is used to check the compatibility of observed and expected frequencies in contingency tables. Suppose there exist k possible outcomes for a random experiment. Denote

Table 14.2: χ^2 test for consistency in $k \times 2$ table

	Class 1	Class 2	Row Totals
Sample 1	x_1	$n_1 - x_1$	n_1
$\vdots$	$\vdots$	$\vdots$	$\vdots$
Sample i	x_i	$n_i - x_i$	n_i
$\vdots$	$\vdots$	$\vdots$	$\vdots$
Sample k	x_k	$n_k - x_k$	n_k
Column Totals	$S = \sum_{j=1}^{k} x_j$	N–S	$N = \sum_{j=1}^{k} n_j$

them by $c_1, c_2, \ldots, c_k$ with respective probabilities $p_1, p_2, \ldots, p_k$. Let n independent trials of the experiment be carried out. Obviously, the closer the observed values are to the expected ones, the better is our measurements. Let np_j denote the expected number of occurrences of c_j, and let f_j be the actual number of occurrences of c_j realized. Note that some of the f_j's can be zero, but $\sum_j f_j = n$. Data on the frequency of occurrence can be conveniently captured into a contingency table. The null hypothesis is that the probability of occurrence follow a known statistical law:

$$H_0 : p_1 = P(c_1), p_2 = P(c_2), \ldots, p_k = P(c_k), \tag{14.30}$$

where $P()$ denotes the assumed distribution (like multinomial law). The observed and expected frequencies form the only two rows in goodness of fit tests, and the number of outcomes (n) is the columns. The same test statistic $\chi_n^2 = \sum_{j=1}^{k}(O_j - E_j)^2/E_j$, which has a χ^2 distribution with $(2-1)(n-1) = (n-1)$ DoF. It is assumed that each of the E_j's are greater than 5. The median test in contingency tables is related in which we estimate the median and then arrange the rows as "below the median" and "above the median" for various populations (columns). Another extension of the goodness-of-fit test is to check if two multinomial distributions with the same number of outcomes are identical or not. In this case the null hypothesis is that $H_0 : p_j = q_j$ for $j = 1, 2, \ldots, k$. If m and n denote the number of random samples drawn from each population, the test statistic is $\chi^2 = \sum_{j=1}^{k}(f_j - m\hat{p}_j)^2/m\hat{p}_j + \sum_{j=1}^{k}(g_j - n\hat{p}_j)^2/n\hat{p}_j$ where $\hat{p}_j = (f_j + g_j)/(m + n)$ is the pooled sample estimate. This has a $\chi^2(k-1)$ distribution.

14.6.5   MINIMUM χ^2-METHOD

As the name implies, this method minimizes the goodness-of-fit statistic in which the expected part contains one or more unknown parameters. Introduced by Smith (1916) [207], it is an alternative to the least squares (LS) and maximum likelihood (ML) methods of estimation. This method might be regarded as a special case of the LS method, as they are asymptotically equivalent (Harsaae (1976) [88]). As the basic assumption in applying the χ^2-method is that each of the observed frequencies are more than 5, this method is not recommended in other cases (although merging of such classes with frequencies less than five will work when there are a large number of them). Nevertheless, it has been applied to select optimal models (like building a decision tree from labeled data by selecting the best discriminating attributes from the root (top level) downward (Chattamvelli (2016) [41]) from among two or more competing models in machine learning, econometrics, etc.

14.6.6 LIKELIHOOD RATIO TESTS

The test statistic for tests of a simple null against a composite alternate hypotheses involve the likelihood ratio $\lambda = L_{max}(\omega)/L_{max}(\Omega)$ where ω is the parameter space confined by H_0, and Ω the parameter space confined by H_a. If the null hypothesis is true, the asymptotic distribution of $-2\log(\lambda)$ is approximately central χ_n^2 where n is the difference between the number of unspecified parameters between H_a and H_0.

Other hypothesis tests that use the χ^2 statistic include the median test for k populations (whether k samples drawn from different populations have the same median), test for multivariate mean vector when variance-covariance matrix is known, Tukey's test of additivity, Cochran–Mantel–Haenszel chi-square test, etc.

14.6.7 χ^2-PLOTS

The squared sample Mahalanobis' distance $D^2 = (\overline{x} - \mu)'S^{-1}(\overline{x} - \mu)$ is asymptotically χ^2-distributed when X deviates from normality. It is essentially a multivariate equivalent of the univariate standardization $(x - \mu)/\sigma$ (or more precisely $(\overline{x} - \mu)/(\sigma/\sqrt{n})$) in which the standard deviation of each variate is used as units along the respective axes. This fact can be used to check for data outliers and test if the parent populations are multivariate normal, and to classify a new data point into one of the many categories. The number of variables p must be the same when D^2 is used to compare two or more groups. It can be geometrically visualized using a χ^2-plot, which is an extension of univariate Q-Q plot, in which the ordered D^2 values are plotted against quantiles of a χ^2 distribution with appropriate DoF. We expect an approximate straight-line segment in the univariate case. There could exist multiple straight-line segments in the multivariate case, and points away from such straight lines are considered to be multivariate outliers (Garrett (1989) [78]). Note, however, that due to correlation among the variables, a data point could turn out to be a multivariate outlier even if it is not a univariate outlier. Mahalanobis' distance can also be used to form clusters from multivariate data by grouping data points that are close together to distinct centroids. This is used in chemometrics for multivariate calibration and process control (De-Maesschalck et al. (2000) [56]). If the centroids (means of distinct populations) are known apriori, we could use it to classify unknown data points.

14.6.8 BIOINFORMATICS AND GENOMICS

Analysis of nucleotide or amino acid sequences are of prime importance in bioinformatics. The χ^2-test is used to analyze group differences when the dependent variable is categorical (nominal or ordinal), which is often the case in these fields. Gene expression data can be cross-tabulated into a $p \times m$ table when p markers and m categories are considered, and into $2 \times m$ table when a single gene with binary traits, or a pair-of genes (as in gene-interaction studies) are considered. Several significance tests in bioinformatics like gene positions in a chromosome vs. panels of molecular markers for various diseases, variations in base sequences adenine (A), guanine (G), cytosine (C), and thymine (T) vs. types of genetic diseases (Mendelian (monogenic), oligogenic (2–4 genes involved), polygenic, etc.), somatic variants vs. feature categories (genomic, evolutionary, consequential), and significance of observed genotype frequencies with Hardy–Weinberg equilibrium use it.

It is used in genomics to compare properties like rate and type (mutation, crossover, inversion, etc.) and their impact on certain gene types (essential genes, positional genes, etc.) to see if they are statistically significant. Deviations, if any, from the expected outcomes like Mendelian ratios in inheritance, alleles in a population, etc. can be quantified using this test procedure. Care must be taken to capture the data as counts or frequencies. A $CT(2 \times n)$ results when one of the categories is binary (differentially enabled genes vs. others, pathway vs. no pathway, etc.). A special case is $CT(2 \times 2)$ that results when both categories are binary (like true positive (TP), true negative (TN), false positive (FP), and false negative (FN) (Shanmugam(2009) [202])). As noted above, the χ^2 test can be misleading when genotype counts are low (say several of them are <5), in which case the Fisher's exact test (FET) is preferred.

14.7 SUMMARY

This chapter introduced the χ^2 distribution and its extensions. Special χ^2 distributions like truncated and generalized forms are more flexible. The chapter ends with a list of applications of χ^2 distribution in inferential statistics, bioinformatics, and genomics.

CHAPTER 15

Student's T Distribution

15.1 INTRODUCTION

This distribution was obtained by William Gosset (1908) [210] as the distribution of the ratio $Z/\sqrt{\chi_n^2/n}$ where $Z \sim N(0, 1)$ and χ_n^2 are independent. It is frequently encountered in small sample statistical inference when population is normally distributed with unknown variance. It is a symmetric, bell-shaped continuous distribution that extends from $-\infty$ to ∞ which has more mass in the shoulders than the Gaussian law.

The Pen-name "Student"

William Gosset was a statistician at Guinness Brewery in Dublin, Ireland. Top management during those days forbade its employees from publishing scientific results because they thought their rival breweries will take advantage of it. That is why Gosset published his work under the pseudonym "Student" in 1908. This distribution was earlier derived as a posterior distribution by Helmert (1876) [91] and Lúroth (1876) [141], and in a more general form by Pearson (1895) [170] as Type IV distribution. This was not the very first paper that Gosset published under *pluma de nombre* Student. Earlier in 1904, he noticed that the number of yeast cells in a random sample of volume suspension of fixed size (constant volume) follows a Poisson distribution. By maintaining homogeneity of yeast cells within fixed limits using the Poisson PMF, the brewery was able to ensure that wines produced during different times were all consistent in taste and quality. He published this result also as "Student," which aroused great interest among scientific community to model 3-D random processes using Poisson law.

It is used in tests for the means, in testing the significance of correlation coefficients, testing significance of regression coefficients, in constructing confidence intervals for means and difference of means, in Bayesian analysis of data from a normal population, etc.

The DoF parameter n is a positive integer in most applications, although theoretically it need not be an integer. As *n* represents the sample size adjusted for "loss of information," it is always an integer in statistical inference. The PDF of classical Student's T (CST) distribution is given by

$$f(t,n) = K(1 + t^2/n)^{-(n+1)/2}, \tag{15.1}$$

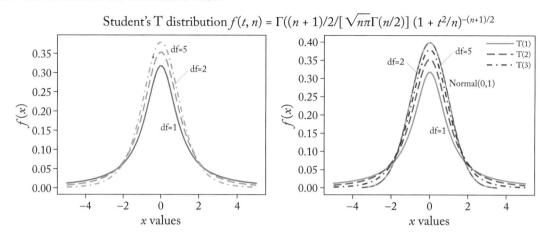

Figure 15.1: T distributions.

where $K = \Gamma((n + 1)/2)/[\sqrt{n\pi}\ \Gamma(n/2)]$. As it is an even function of t, it is always symmetric around $t = 0$ (Figure 15.1).

Alternate representations are

$$f(t;n) = \frac{1}{\sqrt{n}B(\frac{1}{2},\frac{n}{2})}\frac{1}{(1 + t^2/n)^{(n+1)/2}}, \quad n > 0, \tag{15.2}$$

$$f(t;n) = (1/\sqrt{n\pi})\frac{\Gamma((n + 1)/2)}{\Gamma(n/2)}(1 + t^2/n)^{-(n+1)/2}, \quad n > 0, \tag{15.3}$$

and

$$f(t;n) = \left[\sqrt{n}B(\frac{1}{2},\frac{n}{2})(1 + t^2/n)^{(n+1)/2}\right]^{-1}, \quad n > 0. \tag{15.4}$$

We denote it by T_n, $T(n)$ where the DoF (which is usually lowercase English letters like n, m or Greek letter ν) is written either as a subscript or as argument. It is called central T or Student's T (ST) distribution and written as STD(n) or St(n).[1] It finds extensive applications in testing of hypotheses and confidence intervals in the frequentist approach when the population variance is unknown and is estimated from a sample, in economics and finance (portfolio risk management, option pricing, stock returns), among many other fields.

The usual derivation proceeds by first establishing that $\overline{x}$ and s^2 are independent when sampling from a normal population, and then finding the distribution of $t = (\overline{x} - \mu)/(s/\sqrt{n})$. Literally, it is the distribution of the difference between the sample mean and hypothesized (true) mean, divided by the scaled sample standard deviation. The more general form is obtained by a change of

---

[1]Some authors use the lowercase "t" as Student's t distribution for univariate case and T^2 for multivariate case (Hotelling's T^2).

origin and scale transformation $t' = (t - a)/c$. A change of origin and scale transformation results in

$$f(t; a, c, n) = \frac{1}{\sqrt{n} B(\frac{1}{2}, \frac{n}{2})} c/(1 + [(t - a)/c]^2/n)^{(n+1)/2}, \quad n > 0. \tag{15.5}$$

15.2   RELATION TO OTHER DISTRIBUTIONS

It is the distribution of $(\bar{x} - \mu)/(s/\sqrt{n})$ when samples come from a normal population $N(\mu, \sigma^2)$ and σ^2 is unknown. The noncentral T distribution reduces to the central T when the noncentrality parameter is zero. The distribution of T with $n = 1$ is the Cauchy distribution. If X and Y are IID χ^2-distributed random variables with the same DoF, then $(\sqrt{n}/2)(X - Y)/\sqrt{XY}$ is Student's T distributed [30]. If X is an F variate with n DoF, then $T = \frac{\sqrt{n}}{2}(\sqrt{X} - 1/\sqrt{X})$ is Student's T(n). The beta distribution is also related to the T(n) under the transformation $x = 1/(1 + t^2/n)$. The analogue of log-normal to normal distribution is the log-Student's T (LST) distribution

$$f(y, n) = \Gamma\left(\frac{n+1}{2}\right) / \left\{ \sqrt{\pi n} \Gamma(n/2) \, y \left(1 + \frac{1}{n} (\log y)^2\right)^{(n+1)/2} \right\}, \tag{15.6}$$

where $t = \log(y)$ has a Student's T distribution. This distribution is used to model returns from stocks and portfolios. Logarithm of daily returns in stock markets (like DJIA, S&P indices) is approximately ST ($n = 3$) distributed [174]. Cassidy et al. (2010) [33] used a left-truncated LST (n) distribution for this purpose, so that the right-tail decay rate is $O(\exp(-q^2 t^2))$. See Jones (2008) [108] and Li and Nadarajah (2020) [136] for further relationships.

15.3   PROPERTIES OF T DISTRIBUTION

It is always symmetric as it is a ratio distribution with a unit normal in the numerator and the square root of an independent scaled χ^2 distribution in the denominator. Although the χ distribution (square root of χ^2) in the denominator is skewed, it has no impact on the symmetry. The χ distribution approaches symmetry as n becomes large so that the ratio of standard normal to the scaled χ distribution approaches unit-normal law rapidly. It is unimodal with mode $t = 0$. It has a single parameter n, which controls both the spread and peakedness. For higher values of n, the flatness in the tails decreases, and the peakedness increases (Figure 15.1). Eventually it coincides with the standard normal distribution for large n. Write the PDF in (15.1) as $f(t) = K(1 + t^2/n)^{1/2} * (1 + t^2/n)^{-n/2}$. Let $n \to \infty$ and use $\underset{x \to \infty}{Lt}(1 + a/x)^{-x} = e^{-a}$ to get $f(t) = Ke^{-t^2/2}$ as $(1 + t^2/n)^{1/2}$ will tend to one. It can be shown that $K = \Gamma((n + 1)/2)/[\sqrt{n\pi} \, \Gamma(n/2)] \to 1/\sqrt{2\pi}$ as n$\to \infty$. This shows that the limiting distribution is a standard normal. Fujikoshi and Mukaihata (1993) [74] used the transformation $U^2 = (n - .5) \log(1 + t^2/n)$ which converges to normality faster. The distribution is concave upward for $|t| < -\sqrt{(n/(n + 2))}$ and concave downward otherwise. The points of inflections are at $\mp\sqrt{n(n + 2)}$. As in the case of Laplace distribution, random numbers from the T distribution are more prone to producing values that fall far from its mean due to its heavy tails.

15.3.1   MOMENTS AND GENERATING FUNCTIONS

The MGF is easily obtained as

$$M_x(t) = E(e^{tx}) = \int_0^\infty e^{tx} \frac{1}{\sqrt{n}B(\frac{1}{2}, \frac{n}{2})} 1/(1 + x^2/n)^{(n+1)/2} dx \tag{15.7}$$

$$= (\sqrt{n}|t|)^{n/2} \mathbb{K}_{n/2}(\sqrt{n}|t|)/[2^{n/2-1}\Gamma(n/2)], \tag{15.8}$$

where $K_n(.)$ is the modified Bessel function of second kind. The characteristic function is given by

$$\phi(t) = K \int_{-\infty}^\infty e^{itx}/(1 + x^2/n)^{(n+1)/2} dx, \tag{15.9}$$

where we have used the variable x instead of t due to the dummy variable in the ChF (Gaunt (2020) [79]). Upon putting $x^2/n = y^2$ this becomes

$$K\sqrt{n} \int_{-\infty}^\infty \cos(ty\sqrt{n})/(1 + y^2)^{(n+1)/2} dy. \tag{15.10}$$

If n is odd ($= 2m + 1$), this reduces to $\exp(-|t\sqrt{n}|)S_n(|t\sqrt{n}|)$ where S is a polynomial of degree n-1 that satisfies the recurrence $S_{m+3}(t) = S_{m+1}(t) + t^2/(m^2 - 1)S_{m-1}(t)$.

The kth absolute moment is $E(|x|^k) = n^{k/2}\Gamma((k + 1)/2)\Gamma((n - k)/2)/[\sqrt{\pi}\Gamma(n/2)]$. All odd moments (except the first) vanish as this distribution is symmetric. The mean $\mu = 0$ if n>1, and does not exist otherwise. The even moments are given by

$$\mu_k = n^{k/2} \frac{\Gamma(\frac{k+1}{2})\Gamma(\frac{n-k}{2})}{\sqrt{\pi}\Gamma(n/2)}. \tag{15.11}$$

This satisfies the first-order recurrence

$$(n - k)\mu_k = n(k - 1)\mu_{k-2}, \tag{15.12}$$

where μ_0 is assumed to be 1, and $k < n$ is even. The mode is 0 with modal value $(\Gamma(n + 1)/2)/[\sqrt{n\pi}\Gamma(n/2)]$. The variance is $n/(n - 2)$ if $n > 2$. For $n > 4$, the kurtosis coefficient is given by $\beta_2 = 3 + \frac{6}{n-4}$, showing that it is always leptokurtic. Closed-form expressions for central and absolute-central moments (univariate), and absolute moments in multivariate case can be found in Kirkby, Nguyen, and Nguyen (2019) [117]. See Table 15.1 for further properties.

Problem 15.1   If $(x_1, x_2, \ldots, x_n, x_{n+1})$ is a random sample of size $(n + 1)$ from a normal population N $(0, 1)$, find the distribution of $x_{n+1}/\sqrt{(x_1^2 + x_2^2 + \cdots + x_n^2)/n}$.

Problem 15.2   If $Y \sim T_n$, find the distribution of $(1/2) + (1/2) * y/\sqrt{n + y^2}$.

Example 15.3 Mean deviation of Student's T distribution   Find the MD of Student's T distribution.

Table 15.1: Properties of T distribution $(1/[\sqrt{n}\, B(\frac{1}{2}, \frac{n}{2})](1 + \frac{t^2}{n})^{-(n+1)/2})$

Property	Expression	Comments
Range of T	$-\infty < t < \infty$	Infinite
Mean	$\mu = 0$	
Median	0	Mode = 0
Variance	$\sigma^2 = n/(n-2) = 1 + 2/(n-2)$	$n > 2$
Skewness	$\gamma_1 = 0$	Symmetric
Kurtosis	$\beta_2 = 3(n-2)/(n-4) = 3 + 6/(n-4); n > 4$	Always leptokurtic
MD	$\sqrt{n/\pi}\,\Gamma((n-1)/2)/\Gamma(n/2)$	$\int_0^\infty I_{n/(n+t^2)}\left(\frac{n}{2}, \frac{1}{2}\right)dt$
CDF	$F_n(t_0) = 1 - \frac{1}{2}I_{n/(n+t^2)}(n/2, 1/2)$	I = Incomplete beta
Moments	$\mu_k = n^{k/2}\dfrac{\Gamma\left(\frac{k+1}{2}\right)\Gamma\left(\frac{n-k}{2}\right)}{\sqrt{\pi}\Gamma(n/2)}$	$(n-k)\mu_k = n(k-1)\mu_{k-2}$
ChF	$\exp(-|t\sqrt{n}|)S_n(|t\sqrt{n}|)$	

All odd moments except the first vanish, even moments (k even) are given above.

Solution: Let $K = \Gamma((n+1)/2)/[\sqrt{n\pi}\ \Gamma(n/2)]$. As the Student's T distribution is symmetric around zero, the MD is given by

$$MD = K \int_{-\infty}^{\infty} |t|(1 + t^2/n)^{-(n+1)/2}dt. \tag{15.13}$$

Split the integral from $-\infty$ to 0 and 0 to ∞. As $|x| = -x$ when $x < 0$, the first integral becomes $-\int_{-\infty}^{0} t\, f(t)\, dt = \int_0^\infty t\, f(t)\, dt$. Hence,

$$MD = 2K \int_0^\infty t(1 + t^2/n)^{-(n+1)/2}dt. \tag{15.14}$$

Put $t^2 = n\,\tan^2(\theta)$ so that $t = \sqrt{n}\,\tan(\theta)$, and $dt = \sqrt{n}\,\sec^2(\theta)d\theta$. The limits of integration becomes 0 to $\pi/2$ and we get

$$MD = 2Kn \int_0^{\pi/2} \tan(\theta)\,\sec^{-(n+1)}(\theta)\ \sec^2(\theta)d\theta. \tag{15.15}$$

Using $\sec(\theta) = 1/\cos(\theta)$ this becomes

$$MD = 2Kn \int_0^{\pi/2} \sin(\theta)\,\cos^{(n-2)}(\theta)d\theta. \tag{15.16}$$

Put $\cos(\theta) = t$ so that $\sin(\theta)d\theta = -dt$, and the limits are changed as 1 to 0, and we get

$$MD = -2Kn \int_1^0 t^{n-2}dt = 2Kn \int_0^1 t^{n-2}dt. \tag{15.17}$$

As the integral evaluates to $1/(n-1)$, $MD = 2\sqrt{n}/[(n-1)B(1/2, n/2)]$. Expand the complete beta function $B(1/2, n/2) = \Gamma(1/2)\Gamma(n/2)/\Gamma((n+1)/2)$ and write $\Gamma((n+1)/2) = ((n-1)/2)\Gamma((n-1)/2)$. One $(n-1)$ cancels out from numerator and denominator giving the alternative expression $\sqrt{n/\pi}\Gamma((n-1)/2)/\Gamma(n/2)$.

Next, we apply the theorem in Chapter 1 of Part I [47] to find the MD. This gives

$$MD = 2 \int_{ll}^{\mu} F_n(t)dt = 2 \int_{\mu}^{ul} S_n(t)dt = 2 \int_{t=0}^{\infty} S_n(t)dt. \tag{15.18}$$

As the SF can be expressed in terms of the IBF as

$$1 - S_n(t) = F_n(t) = \frac{1}{2}\left[1 + \text{sign(t)} I_y(1/2, n/2)\right], \tag{15.19}$$

where $y = t^2/(n + t^2)$, and $\text{sign}(t) = -1$ for $t < 0$ so that (15.19) becomes

$$MD = 2 \int_0^{\infty} \frac{1}{2}\left[1 - I_y(1/2, n/2)\right]dt. \tag{15.20}$$

Using $1 - I_y(1/2, n/2) = I_{1-y}(n/2, 1/2)$ where $1 - y = n/(n + t^2)$, this becomes

$$MD = \int_0^{\infty} I_{1-y}(n/2, 1/2)dt. \tag{15.21}$$

To evaluate this integral, take $u = I_{1-y}(n/2, 1/2)$ and $dv = dt$ so that $v = t$. Use the chain rule of differentiation to get

$$du = (\partial/\partial t)I_{1-y}(n/2, 1/2) = (\partial/\partial y)I_{1-y}(n/2, 1/2) * (\partial y/\partial t). \tag{15.22}$$

Differentiate $1 - y = n/(n + t^2)$ to get $-\partial y/\partial t = -2nt/(n + t^2)^2$. Also, $\frac{\partial}{\partial y}I_{1-y}(n/2, 1/2) = g_{1-y}(n/2, 1/2)$ where g() is the PDF of BETA-I. Integrate (15.21) by parts to get

$$t * \left[I_{1-y}(n/2, 1/2)\right] \Big|_0^{\infty} + \int_0^{\infty} t * g_{1-y}(n/2, 1/2)(2nt/(n + t^2)^2)dt. \tag{15.23}$$

The first term is zero using L'Hospital's rule. Take $2n$ outside the integral to get

$$MD = 2n \int_0^{\infty} [t^2/(n + t^2)^2] * g_{1-y}(n/2, 1/2)dt. \tag{15.24}$$

Put $v = n/(n + t^2)$, and $1 - v = t^2/(n + t^2)$. This gives $t = \sqrt{n}((1 - v)/v)^{1/2}$, and $dv = -2nt/(n + t^2)^2 dt$. Write $[t^2/(n + t^2)^2] = [t^2/(n + t^2)] * 1/(n + t^2) = v(1 - v)/n$, and $dv = (-2/\sqrt{n})v^{1/2}(1 - v)^{3/2}$. Expand $g_{1-y}(n/2, 1/2)$. The n cancels out from the numerator and denominator, and (15.24) becomes

$$(\sqrt{n}/B(n/2, 1/2)) \int_0^1 v^{(n-3)/2}(1 - v)^0 dv. \tag{15.25}$$

This simplifies to $(\sqrt{n}/B(n/2, 1/2))/[(n - 1)/2] = 2\sqrt{n}/[(n - 1) * B(n/2, 1/2)]$ which is the same expression obtained before. Expand $B(n/2, 1/2)$, then write $\Gamma((n + 1)/2)$ as $[(n - 1)/2]\Gamma((n - 1)/2)$, and cancel out 2 and $(n - 1)$ to get the expression given in Table 15.1.

15.3.2   TAIL AREAS

The CDF of a Student's t random variable is encountered frequently in small sample statistical inference. For example, it is used in tests for the means, in testing the significance of correlation coefficients, and in constructing confidence intervals for means. The CDF is given by

$$F(x; n) = \frac{1}{2} + x\Gamma((n + 1)/2)_2F_1(\frac{1}{2}, \frac{n + 1}{2}, \frac{3}{2}; -x^2/n)/(\sqrt{n\pi}\Gamma(n/2)), \tag{15.26}$$

where $_2F_1(a, b, c; x)$ is Gauss hypergeometric function.

The area under ST (n) from $-t$ to $+t$ is of special interest in finding two-sided confidence intervals and tests of hypothesis. We denote it as $T(-t : t|n)$ or $T_n(-t : t)$.

$$T_n(-t : t) = \frac{1}{\sqrt{n}B(\frac{1}{2}, \frac{n}{2})} \int_{-t}^t (1 + x^2/n)^{-(n+1)/2} dx. \tag{15.27}$$

This integral can be converted to an IBF by the transformation $y = n/(n + x^2)$ giving

$$T_n(-t : t) = 1 - I_{n/(n+t^2)}(n/2, 1/2). \tag{15.28}$$

Due to the symmetry, the area from $\pm t$ to the mode $(x = 0)$ is

$$T_n(-t : 0) = T_n(0 : t) = \frac{1}{2} - \frac{1}{2}I_{n/(n+t^2)}(n/2, 1/2). \tag{15.29}$$

The CDF (area from $-\infty$ to $t > 0$) is given by

$$F_n(t_0) = \int_{-\infty}^{t_0} f(t)dt = \int_{-\infty}^0 f(t)dt + \int_0^{t_0} f(t)dt. \tag{15.30}$$

Due to symmetry, the first integral evaluates to 1/2. Represent the second integral using (15.29) to get

$$F_n(t_0) = 1 - \frac{1}{2}I_y(n/2, 1/2), \tag{15.31}$$

where $y = n/(n + t_0^2)$, and $I_x(a, b)$ is the IBF.

For even degrees of freedom, the CDF of Student's T distribution can be obtained as

$$F_n(t) = \frac{1}{2}(1 + \sqrt{(x/\pi)} \sum_{i=0}^{\frac{n}{2}-1} (1 - x)^i \Gamma(i + 1/2)/\Gamma(i + 1)), \text{ where } x = t^2/(n + t^2). \quad (15.32)$$

The special cases $n = 2$ and $n = 4$ are $F_2(t) = \frac{1}{2}\left(1 + t/\sqrt{(2 + t^2)}\right)$ and $F_4(t) = \frac{1}{2}\left[1 + (1 + 2/(4 + t^2))t/\sqrt{(4 + t^2)}\right]$. As mentioned above, this reduces to Cauchy CDF for $n = 1$ as $\frac{1}{2} + \frac{1}{\pi}\text{sign}(t) \tan^{-1}(t)$. For $n = 3, 5$ similar expressions exist (see Balakrishnan and Nevzorov (2003) [15]). See Tiku (1971) [214] for the T distribution under non-normal situations, Good and Smith (1986) [81] for a power series for the SF, and Singh and Lee (1988) [206] for an algorithm for the CDF.

15.3.3 APPROXIMATIONS

The standard normal distribution $Z(0, 1)$ provides a good approximation to the T-distribution when DoF > 30. Note that the tails of Gaussian law decreases exponentially whereas that of T distribution decreases polynomially, so that approximations in extreme tails may not be accurate for small DoF. Fisher (1925) [66] gave an approximation when n is large. Finner, Dickhaus, and Roters (2008) [65] obtained highly accurate approximations to tail areas using the ratio of T and N (0, 1) variates.

15.4 RANDOM NUMBERS

There are many ways to generate random numbers from this distribution. One of the most efficient is the polar algorithm (Bailey (1992) [11]) that uses two uniform random numbers.

Algorithm 15.1 Random numbers from Student's t distribution

1: Generate U[0,1] random variates x and y
2: set count =1
3: while count <= n
4:     Find u = 2x-1, v=2y-1
5:     Find w = $u^2 + v^2$
6:     If w$\leq$1 then
7:         t = u $\sqrt{n} * (1/w^{2/n} - 1)/w$
8:         count = count+1; Generate new x,y from U[0,1]
9:     endif

The *count* is incremented only when a valid random number has been generated. The generated number (t) can be stored in an array or list for further processing. The inverse CDF method can be used to directly generate random numbers for special values of DoF (like $n = 1, 2, 4$ in which case exact closed form solutions exist).

15.5   EXTENSIONS OF T DISTRIBUTIONS

A direct extension of T distribution is the noncentral T distribution which can arise in two situations. Consider the definition of central T as $T = Z/\sqrt{\chi_n^2/n}$ where Z is the standard normal, and χ_n^2 is a central chi-squared random variable. When the Z in the numerator has a non-zero mean, the distribution of T is the classical noncentral T (NCT). When the χ_n^2 in the denominator is a noncentral, we get type-II noncentral T distribution. Combining both cases above (Z has nonzero mean and $\chi_n^2(\lambda)$ is noncentral) results in doubly noncentral T distribution (Chattamvelli (1995) [36], Kocherlakota and Kocherlakota (1991) [118]).

McDonald and Newey's Student's T (MNST) distribution [148] has PDF

$$f(t; n, a) = K(n, a)(1 + |t|^a/n)^{-n-1/a}, \tag{15.33}$$

where $K(n, a) = a/[2n^{1/a} B(1/a, n)]$. Integrate wrt t from 0 to x and use the symmetry property to get the CDF as

$$F(x; n, a) = 0.50(1 + \text{sign}(x) I_y(1/a, n)), \tag{15.34}$$

where $y = |t|^a/(n + |t|^a)$ and $I_y(a, b)$ is the incomplete beta function. Another extension of Student's T distribution due to Harvey and Lange (HLST) (2015) [90] has PDF

$$f(t; n, a) = K(n, a)(1 + |t|^a/n)^{-(n+1)/a}, \tag{15.35}$$

where $K(n, a) = a/[2n^{1/a} B(n/a, 1/a)]$. The kth absolute moment of this distribution is $E(|x|^k) = n^{k/a}\Gamma((k+1)/a)\Gamma((n-k)/a)/[\Gamma(1/a)\Gamma(n/a)]$ for $0 < k < n$. Note that the shape parameter a need not be an integer, but it reduces to the CST when $a = 2$. A change of origin and scale transformation can be applied to the above forms to get distributions with additional parameters.

A transmuted T distribution (TTD) can be obtained with the CDF

$$F(x; a, b) = (1 + \lambda)G(x) - \lambda[G(x)]^2, \quad -\infty < x < \infty, \ -\infty < a < \infty, \ b > 0, |\lambda| \le 1, \tag{15.36}$$

where $G(x)$ is the CDF of one of the T distributions.

15.5.1   FOLDED T DISTRIBUTION

As the CST is symmetric around zero, we could get the folded Student's T distribution (FSTD), which is the distribution of $Y = |T|$ as

$$f(y; n) = \frac{2}{\sqrt{n} B(\frac{1}{2}, \frac{n}{2})}(1 + y^2/n)^{-(n+1)/2}, \quad n, y > 0. \tag{15.37}$$

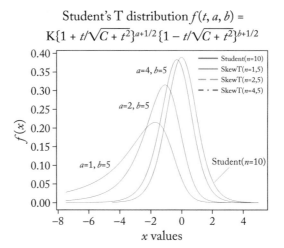

Figure 15.2: Skew T distributions.

This reduces to the folded Cauchy distribution for $n = 1$. The mean and variance are $\mu = 2\sqrt{n/\pi}\Gamma((n+1)/2)/[(n-1)\Gamma(n/2)]$ for $n > 1$ and $\sigma^2 = (n/(n-2) - 4n(\Gamma((n+1)/2))^2/[\pi(n-1)^2](\Gamma(n/2))^2)$ for $n > 2$. As the range of this distribution is $(0,\infty)$ a log-FSTD distribution can be obtained as $Y = \log(X)$. This has applications in risk management, insurance loss, among many other fields (Brazauskas and Kleefeld (2011) [28]). A review of generalizations and software available for them can be found in Li and Nadarajah (2018) [136].

15.5.2   SKEW T DISTRIBUTION

By factoring the PDF into two asymmetric products, Jones and Faddy (2003) [111] obtained the skew-t distribution

$$f(t;a,b) = K\left\{1 + t/\sqrt{C+t^2}\right\}^{a+\frac{1}{2}}\left\{1 - t/\sqrt{C+t^2}\right\}^{b+\frac{1}{2}}, \tag{15.38}$$

where K is given by $1/K = C^{1/2}2^{C-1}\,B\,(a,b)$ and $C = (a+b)$. This reduces to to ST $(2a)$ when $a = b$. The rth raw moment is $E(X^r) = C^{r/2}/[2^r\,B(a,b)]\sum_{j=0}^{r}(-1)^j\binom{r}{j}B(a+r/2-j,b-r/2+j)$ for $r < \min(2a, 2b)$. See Figure 15.2.

15.6   APPLICATIONS

This distribution finds extensive applications in financial risk modeling, and statistical inference, especially in tests of hypotheses and confidence intervals (Kanji (2006) [112]). It is used in those situations when the population is normally distributed, the sample size is small (say < 20), but the variance is unknown. The original work of William Gosset ("Student") involved such extremely small samples with $n \leq 5$ (see also Lehmann (2012) [127], de Winter (2013) [54]). As mentioned in the last chapter, the distribution of LRT is $\sim \chi_n^2$ asymptotically. As the normal and chi-square approximations for LRT are valid for large sample sizes only, it is preferable to use the t distribution for $n < 30$. It is used not only in linear regression models, but in nonlinear regression and mixed effects models.

Student's T filters (STF) that assume the latent state and background noise to have a joint T distribution is popular in DSP (Tronarp and Karvonen (2019) [217]). Gain calibration algorithms for radio interferometric antennas based on heavy-tailed distributions like $T(n)$ are less suscepti-ble to calibration artifacts, and can mitigate biases introduced by incomplete sky models and radio frequency interference due to their robustness against outliers in the data (Sob et al. (2020) [208]). Meitz, Preve, Saikkonen (2018) [151] used a mixture autoregressive model based on Student's t-distribution in volatility forecasting. Gao, Wen, and Wang (2017) [76] used ST mixture model (STMM) to segment two-phase images using active contours method. Some specific applications in statistical inference appears in the following paragraphs. It is assumed that the test statistic has a T distribution under the null hypothesis H_0 in all of the tests discussed below.

15.6.1 T-TEST FOR MEANS

Perhaps the most popular use of this distribution is to test the mean of a normal population with unknown variance when the sample size is small (say < 25). The unknown variance is replaced by the sample variance and the test statistic becomes $(\bar{x} - \mu_0)/(s/\sqrt{n})$, where $\bar{x}$ is the sample mean and s is the sample standard deviation. This has a $T(n-1)$ distribution. A $100(1-\alpha)$% CI for μ can then be developed as $(\bar{x} \mp t(n-1; \alpha/2) s/\sqrt{n})$. The T-tests for difference in means involve independent samples (unpaired samples) or paired samples. The $100(1-\alpha)$% CI for $\mu_x - \mu_y$ can similarly be developed as $(\bar{x} - \bar{y}) \mp t(m; \alpha/2)\sqrt{s_x^2/n_x + s_y^2/n_y}$, where the DoF m is estimated from the nearest integer of $(s_x^2/n_x + s_y^2/n_y)^2/[(s_x^2/n_x)^2/(n_x - 1) + (s_y^2/n_y)^2/(n_y - 1)]$.

15.6.2 T-TEST FOR LINEAR REGRESSION

Several tests of hypotheses in linear regression use the CST. For example, test for the significance of the regression coefficient of y on x uses $T(n-2)$ distribution. Confidence intervals (CI) for regres-sion coefficients also use $T(n-2)$ distribution. Thus, $100(1-\alpha)$% CI for simple linear regression slope β_0 is given by $\hat{\beta}_0 \mp t(n-2, \alpha/2)s_{\beta_0}$ where $s_{\beta_0} = s\sqrt{1/n + \bar{x}^2/\sum_k (x_k - \bar{x})^2}$. Similarly, inference on mean response (as well as CI for it) $\hat{y} = \hat{\beta}_0 + \hat{\beta}_1 x$ use $T(n-2)$ distribution.

15.6.3 T-TEST FOR CORRELATION

This test is used to check if a sample (x_k, y_k) for $k = 1, 2, \ldots, n$ has been drawn from a bivariate normal population with $\rho = 0$. The test statistic is $t = \sqrt{n-2}\, r/\sqrt{1-r^2}$ which has a $T(n-2)$

distribution. It is implicitly assumed that the relationship, if any, is linear. If our alternate hypothesis is $\rho \neq 0$, it is a two-tailed test; otherwise, it is one-tailed.

15.7   SUMMARY

This chapter introduced the T distribution and its extensions. Most important properties of the distribution are summarized. Special T distributions like truncated, folded, skewed, and log-T distributions are briefly introduced. The chapter ends with a list of applications of T distribution in statistical inference.

CHAPTER 16

F **Distribution**

16.1 INTRODUCTION

This distribution was obtained by Fisher (1908) as the distribution of the ratio $F = (\chi^2(m)/m)/(\chi^2(n)/n) = \frac{n}{m}(\chi^2_m/\chi^2_n)$ where the numerator and denominator are independent. It is frequently encountered in small sample statistical inference as the null distribution of a test statistic in the analysis of variance (ANOVA). It is a continuous distribution with PDF

$$f(x;m,n) = \frac{\Gamma((m+n)/2)m^{m/2}n^{n/2}}{\Gamma(m/2)\Gamma(n/2)} \frac{x^{m/2-1}}{(n+mx)^{(m+n)/2}}, \quad 0 < x < \infty. \tag{16.1}$$

It is also called Snedecor's F distribution or the Fisher-Snedecor F distribution (Snedecor (1934) gave the name "F distribution" in honor of Fisher and tabulated it widely). It is used extensively in ANOVA and related procedures as shown in the applications section. This is due to the normality assumption of the population from which samples come, so that the null distribution of the test statistic has an F distribution. It is also used in computing the power of various statistical tests that use the sample variance.

The DoF parameters m and n are positive integers in most applications, although theoretically they need not be integers. The first parameter m corresponds to the numerator χ^2 distribution and is called numerator DoF, and the second one is the denominator DoF. Exchanging the numerator and denominator variates results in another F distribution. As the shape of the distribution depends on the DoF parameters, it is important to distinguish between them while referring F tables.

16.1.1 ALTERNATE FORMS

Alternate representations include

$$f(x;m,n) = (m/n)^{m/2}x^{m/2-1}/[B(m/2,n/2)(1+mx/n)^{(m+n)/2}], \tag{16.2}$$

$$f(x;m,n) = [m^{m/2}n^{n/2}/B(m/2,n/2)]\, x^{m/2-1}\,(n+mx/n)^{-(m+n)/2}, \tag{16.3}$$

and

$$f(x;m,n) = \sqrt{(mx)^m n^n/(n+mx)^{(m+n)}}/[x\,B(m/2,n/2)]. \tag{16.4}$$

By writing $1+mx/n = (n+mx)/n$ and separating out $1/2$ in the exponent, (16.2) can be written as

$$f(x;m,n) = \left((mx)^m n^n/(mx+n)^{m+n}\right)^{1/2}/[xB(m/2,n/2)]. \tag{16.5}$$

Several textbooks and research papers use ν_1 and ν_2 as numerator and denominator DoF (in place of m and n). We denote the central F distribution by $F_{m,n}$ or $F(m, n)$ where the first argument (m) is always the numerator DoF. Simple reduction formula exist to find the PDF and CDF when either m or n is even, as shown later.

16.2 RELATION TO OTHER DISTRIBUTIONS

The noncentral F distribution reduces to the central F when the noncentrality parameter is zero. If $X \sim F(m, n)$, then $Y = \lim_{n \to \infty} mX$ is central χ^2_m distributed. If $X \sim F(m, n)$ then $Y = 1/X$ is $F(n, m)$. This is the reason why the F-tables are almost always prepared as one-tail tables. As the T distribution is the ratio of a standard normal to the square root of an independent scaled χ^2_n random variate, the square of T is F distributed with 1 and n DoF. If X and Y are independent F variates with the same DoF, then $T = \frac{\sqrt{n}}{2}(\sqrt{X} - \sqrt{Y})$ is Student's $T(n)$ (Cacoullos (1964) [30]). A scaled Hotelling T^2 variate is also related to the F distribution as $nT^2(m, N)/(mN) \sim F(m, n)$ where $N = (m + n - 1)$. Tail area of binomial distribution is related to the F distribution as

$$\sum_{x=0}^{k} \binom{n}{x} p^x q^{n-x} = 1 - F_y(2(k + 1), 2(n - k)) \text{ where } y = p(n - k)/(q(1 + k)). \quad (16.6)$$

The CDF of central F distribution in terms of the IBF is

$$F_{m,n}(x) = I_y(m/2, n/2), \quad (16.7)$$

where (m, n) are the numerator and denominator DoF and $y = mx/(n + mx)$. If X_1 and X_2 are IID Laplace(a, b) random variables, then $Y = |X_1/X_2|$ has an $F(2, 2)$ distribution. If X_1 and X_2 are IID GAMMA (a_i, b_i), $i = 1, 2$ random variables, then $Y = (X_1/X_2)$ has a scaled F distribution. More precisely, $k(X_1/X_2) \sim F(2a_1, 2a_2)$ where $k = a_2 b_1/(a_1 b_2)$. If $X \sim F(m, n)$ then $Y = (mX/n)/[1 + (mX/n)] \sim$ Beta-I $(m/2, n/2)$, and $mX/n \sim$ Beta-II $(m/2, n/2)$. Fisher's Z distribution is also related to the F distribution as $Z = \frac{1}{2}\log(F)$ (page 231).

16.3 PROPERTIES OF *F* DISTRIBUTION

As both the numerator and denominator variates in the definition are χ^2, this distribution is defined for $x > 0$. Due to symmetry in the definition as a ratio, $Y = 1/F$ has exactly identical distribution with the degrees of freedom reversed. The distribution of $Z = (1/2)\log(F)$ is more tractable, as it converges to normality faster than F itself. As the χ^2 distribution is a special case of gamma distribution, the ratio of two properly scaled independent gamma variates has an F distribution. The F distribution has a long right tail, and is skewed to the right for small parameter values (Figure 16.1). The unscaled F distribution is the distribution of the ratio $\chi^2(m)/\chi^2(n)$, which is BETA-II $(m/2, n/2)$.

F distribution $f(x; m, n) = \Gamma((m + n)/2)m^{m/2}n^{n/2}/\Gamma(m/2)\Gamma(n/2)x^{m/2-1}/(n + mx)^{(m+n)/2}$

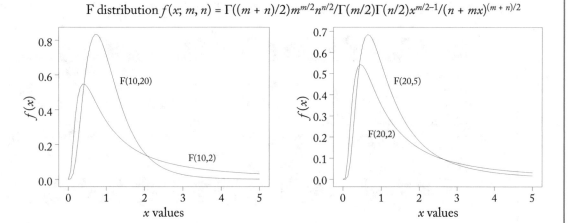

Figure 16.1: F distributions.

16.3.1 MOMENTS AND GENERATING FUNCTIONS

The mean is undefined when $n < 2$, but it is $n/(n-2) = 1 + 2/(n-2)$ for $n > 2$. This does not depend on the numerator DoF parameter m. Although the distribution has infinite range, the mean (center of mass) is bounded by 3 (for integer values of n), and rapidly approaches 1 as n becomes large. The kth raw moment is given by

$$\mu_k = (n/m)^k \Gamma(m/2 + k)\Gamma(n/2 - k)/[\Gamma(m/2)\Gamma(n/2)], \tag{16.8}$$

for $k < n/2$. The variance is $\sigma^2 = \frac{2n^2(m+n-2)}{m(n-2)^2(n-4)}$, which is defined for $n > 4$. This in terms of the mean is $2\mu^2/(n-4) * ((m+n-2)/m)$. The mode is $[(m-2)/m] * [n/(n+2)]$. As $n/(n-2)$ is > 1 and $n/(n+2)$ is < 1, the mode is less than the mean. As n becomes large, the mean tends to 1 but the mode tends to $(m-2)/m$. Similarly, the skewness coefficient is undefined for $n \leq 6$ (all of these conditions are on n and not on m). For $n > 6$, the skewness coefficient is $\beta_1 = \frac{2(2m+n-2)\sqrt{2(n-4)}}{\sqrt{m}(n-6)\sqrt{m+n-2}}$.

The characteristic function of F variate is (Phillips (1982) [173])

$$\phi(t; m, n) = \Gamma((m + n)/2)/\Gamma(n/2)\psi(m/2, 1 - n/2, -itn/m), \tag{16.9}$$

where $\psi(m/2, 1 - n/2, -itn/m)$ is the confluent hypergeometric function of type-2. A double infinite sum for it is as follows ([9], [172]):

$$\phi(m, n; t) = \frac{1}{B(m/2, n/2)} \sum_{i=0}^{\infty} \sum_{j=0}^{\infty} \frac{(it)^i}{[i!(i + j + m/2)]} \binom{i + j - n/2}{j}, \tag{16.10}$$

which is valid for n even. See Table 16.1 for further properties.

Table 16.1: Properties of F distribution ($\frac{\Gamma((m+n)/2)m^{m/2}n^{n/2}}{\Gamma(m/2)\Gamma(n/2)}\frac{x^{m/2-1}}{(n+mx)^{(m+n)/2}}$)

Property	Expression	Comments
Range	$0 \le x < \infty$	Infinite
Mean	$\mu = n/(n-2)$	$= 1 + 2/(n-2)$
Mode	$n(m-2)/[m(n+2)]$	$m > 2$
Variance	$\sigma^2 = 2\mu^2(m+n-2)/[m(n-4)]$	$\mu > \sigma^2$
CV	$(2(m+n-2)/[m(n-4)])^{1/2}$	
Skewness	$(2m+n-2)[8(n-4)]^{1/2}/[\sqrt{m}(n-6)(n+m-2)^{1/2}]$	γ_1
Kurtosis	$\beta_2 > 3$	
MD	$E\lvert X-\mu\rvert = 2\int_0^{n/(n-2)} I_{mx/(n+mx)}(m/2, n/2)dx$	
Moments	$(n/m)^r\Gamma(m/2+r)\Gamma(n/2-r)/[\Gamma(m/2)\Gamma(n/2)]$	μ_r'
ChF	$[\Gamma((m+n)/2)/\Gamma(n/2)] \Psi(m/2, 1-n/2; -nit/m)$	
Additivity	$\sum_{i=1}^m F(m_i, n) = F(\sum_{i=1}^m m_i, n)$	IID F
Tail area	$I_{n/(n+mx)}(n/2, m/2) = 1 - I_{mx/(n+mx)}(m/2, n/2)$	$\Pr(F \le x)$

Additivity is for unscaled F with the same denominator DoF.

16.3.2   MOMENT RECURRENCES

Consider the density $g_{m,n}$ of m/n times an $F_{m,n}$ random variable, namely $g_{m,n}(y) = y^{m/2-1}/[B(m/2, n/2)(1+y)^{(m+n)/2}]$. The basic density recurrence is

$$m(1+y)g_{m+2,n}(y) = (m+n)yg_{m,n}(y), \tag{16.11}$$

a weighting of $g_{m,n}$ by $y/(1+y)$. The relationship between central density and distribution functions is $n[G_{m,n+2}(y) - G_{m-2,n+2}(y)] = -2g_{m,n}(y)$. Write $\mu_{m,n}^k = E\left[\{\chi_m^2(\lambda)/\chi_n^2\}^k\right]$ (and note that $E[F_{m,n}^k] = (n/m)^k\mu_{m,n}^k$). The moment recurrence is $n\mu_{m,n+2}^{k+1} = m\mu_{m+2,n}^k$. Another density recurrence is

$$ng_{m,n+2}(y) = (m+n)g_{m,n}(y) - mg_{m+2,n}(y), \tag{16.12}$$

from which a moment recurrence follows as $n\mu_{m,n+2} = (m+n)\mu_{m,n} - m\mu_{m+2,n}$. Several recurrences satisfied by the density, distribution function, and moments can be found in Chattamvelli and Jones (1995) [42].

Example 16.1   Mean deviation of *F* distribution   Find the mean deviation of *F* distribution.

Solution: The MD is given by

$$MD = 2 \int_{ll}^{\mu} F_n(t)dt = 2 \int_{\mu}^{ul} S_n(t)dt = 2 \int_{x=0}^{\infty} S_n(t)dt. \qquad (16.13)$$

From this we get the MD as

$$MD = 2 \int_0^{\mu} I_{n/(n+mx)}(n/2, m/2)dx. \qquad (16.14)$$

This can be written in terms of beta integral by taking $u = I_{n/(n+mx)}(n/2, m/2)$ and $dv = dx$, as done in Chapter 13. •

16.3.3   APPROXIMATIONS

An $O(1/n^{3/2})$ approximation to the CDF is $F(x; m, n) = \Pr[F \leq x] = \Phi(r - \log(r/q)/r)$ where $r = \text{sign}(x - 1) \left((m + n)\log((mx + n)/(m + n)) - m\log(x)\right)^{1/2}$, and $q = (x - 1)(mn(m + n)/2)^{1/2}/(mx + n)$ (Wong (2008) [232]). Li and Martin (2002) [135] approximated the tail areas using the χ^2 distribution, which was extended by Jiang and Wong (2018) [102], who obtained highly accurate approximations via an adjusted log-likelihood ratio statistic.

16.4   TAIL AREAS

The CDF of an F random variable is encountered frequently in small sample statistical inference. For example, it is used in tests for equality of several means, equality of two variances from samples drawn from normal populations, in testing the significance of regression coefficients, and in constructing confidence intervals for ratio of variances.

Integrating from 0 to $+x$ gives the CDF of a Snedecor's F distribution with (m, n) degrees of freedom as

$$F_{(X)}(x; m, n) = I_{n/(n+mx)}(n/2, m/2) = 1 - I_{mx/(n+mx)}(m/2, n/2). \qquad (16.15)$$

The tail areas are related as $F(x; m, n) = 1/F(1 - x; n, m)$. The special cases are $F(x; 1, 1) = (\frac{2}{\pi})\tan^{-1}(\sqrt{mx/n})$, $F(x; 1, 2) = \sqrt{mx/(n + mx)}$, $F(x; 2, 1) = 1 - \frac{\sqrt{n}}{\sqrt{(n+mx)}}$, $F(x; 2, 2) = mx/(n + mx)$. Barabesi and Greco (2002) [17] obtained closed-form expressions to evaluate the CDF using the above relationships.

Problem 16.2   Prove that the mode of an F distribution is $((m - 2)/m)(n/(n + 2))$ for $m > 2$.

16.5   EXTENSIONS OF F DISTRIBUTION

A direct extension of F distribution is the noncentral F distribution which can arise in two situations. Consider the definition of central F as $F = (\chi_m^2/m)/(\chi_n^2/n)$ where the numerator and denominator are independent. When the χ_m^2 in the numerator is noncentral, the distribution of F is the classical noncentral F (NCF). When the χ_n^2 in the denominator is a noncentral, we get type-II noncentral F distribution. Combining both cases above results in doubly noncentral F (DNF) distribution (Kocherlakota and Kocherlakota (1991) [118], Chattamvelli (1995) [36]). Computationally efficient polynomial algorithms for NCF and DNF appear in Chattamvelli (1996) [38].

16.5.1   SIZE-BIASED *F* DISTRIBUTIONS

As the mean of an F distribution is $n/(n-2)$, a size-biased F distribution can be obtained as

$$f(x; n) = (1 - 2/n) \frac{\Gamma((m+n)/2) m^{m/2} n^{n/2}}{\Gamma(m/2) \Gamma(n/2)} \frac{x^{m/2}}{(n+mx)^{(m+n)/2}}. \tag{16.16}$$

Alternately, consider the linear combination $(1 + cx)$ with expected value $((1+c)n - 2)/(n-2)$ and proceed as in Chapter 1 of Part I.

16.6   FISHER'S *Z* DISTRIBUTION

This distribution is obtained as a transformation from the F distribution as $Z = \frac{1}{2} \log(F)$. It is also called logarithmic F distribution. As the range of F is from 0 to ∞, the range of Z is $-\infty$ to ∞. The PDF is obtained directly as

$$f_z(m, n) = \frac{2m^{m/2} n^{n/2}}{B(m/2, n/2)} \frac{e^{mz}}{(n + me^{2z})^{(m+n)/2}}, \tag{16.17}$$

where $B(m/2, n/2)$ is the CBF.

The unnormalized Z distribution results when F in $Z = \frac{1}{2} \log(F)$ is replaced by the unnormalized F (which is BETA-II). The corresponding Z is singly noncentral when the F-distribution is noncentral. If both chi-squares in the F-distribution are noncentral, the corresponding Z is called doubly noncentral Z ([38], [197]).

16.6.1   PROPERTIES OF FISHER'S *Z* DISTRIBUTION

Left tail areas can be expressed in terms of incomplete beta function as follows:

$$Z_{m,n}(x) = P[\frac{1}{2} \log(F_{m,n}(x)] = P[F_{m,n}(x) \leq e^{2x}] = 1 - I_c(m/2, n/2), \tag{16.18}$$

where $c = me^{2x}/(n + me^{2x})$. A symmetry relationship connecting tail areas of $Z(x; m, n)$ and $Z(x; n, m)$ appears in Chattamvelli (1995) [36] as $Z_c(m, n) = 1 - Z_{-c}(n, m)$. See Table 16.2 for further properties.

16.6.2   MOMENTS

The characteristic function is $(n/m)^{it/2} \Gamma((n-it)/2) \Gamma((m+it)/2)/[\Gamma(m/2) \Gamma(n/2)]$. Using the derivatives of gamma function, the first two moments are $\mu = (m-n)/[2mn] = (1/n - 1/m)/2$ and

Table 16.2: Properties of Fisher's Z $\left(\frac{2m^{m/2}n^{n/2}}{B(m/2,n/2)}\frac{e^{mz}}{(n+me^{2z})^{(m+n)/2}}\right)$

Property	Expression	Comments
Range of Z	$-\infty \leq z \leq \infty$	Infinite
Mean	$\mu \simeq (m-n)/(2mn)$	Mode $z = 0$
Variance	$\sigma^2 \simeq (m+n)/(2mn)$	$\Rightarrow \mu < \sigma^2$
ChF	$(n/m)^{it/2}\dfrac{\Gamma((n-it)/2)\Gamma((m+it)/2)}{[\Gamma(m/2)\Gamma(n/2)]}$	
Tail area	$I_c(m/2, n/2), c = me^{2x}/(n + me^{2x})$	
Symmetry relation	$Z(x; m, n) = 1 - Z(-x; n, m)$	

$\mu_2 = (m+n)/[2\ mn] = (1/n + 1/m)/2$ approximately. The cumulants are easier to find in terms of digamma function [7], [230].

16.6.3   RELATION TO OTHER DISTRIBUTIONS

When both of the parameters $\to \infty$, $Z \to N(\frac{1}{2}\frac{m-n}{mn}, \frac{1}{2}\frac{m+n}{mn})$. Convergence of Z to normality is faster than the convergence of F distribution. If $X \sim Z(m,n)$ then $\exp(2Z) \sim F(m,n)$. The transformation $V = (N/(N+1))^{1/2}(Z/b)$ is approximately distributed as T_N where $N = m + n - 1$, $b^2 = \frac{1}{2}(1/m + 1/n)$ and T_N is Student's T distribution.

16.7   APPLICATIONS

The F distribution finds extensive applications in statistical inference, especially in tests of hypotheses and confidence intervals. It is used in testing the equality of variances of two IID normal populations, equality of means in one-way ANOVA, testing the significance of a normal linear regression model, tests for equality of two sets of regression coefficients, tests for linear constraints on regression coefficients, tests for lack of fit in linear regression, tests for discriminant functions (Fisher (1940)) [68], Rao's U-test for additional discrimination, etc. Wilks λ criterion used in multivariate analysis has an F distribution for special values of the parameters. Transformations of λ criterion are used in exact right-sided tests. The most popular transformations are of the form $T_1 = C_1(1-\lambda)/\lambda$ (Wilks (1932) [228]) and $T_2 = C_2(1-\sqrt{\lambda})/\sqrt{\lambda}$ (Rao's F-test) where $C_1 = (n_e + n_h - p)/p$, $C_2 = (n_e + n_h - p - 1)/p$ are functions of degrees of freedom associated with hypothesis n_h, error term n_e, and dimensionality (number of the variables) p. Then $T_1 \sim F(p, (n_e + n_h - p))$ and $T_2 \sim F(2p, 2(n_e + n_h - p - 1))$ both have central F distribution.

   If the error terms in a regression model are assumed to be $N(0, \sigma^2 I_n)$ distributed, the test for significance of the entire model can be based on the F-statistic as $F = (\sum_k (\hat{y}_k - \bar{y})^2/p)/(\sum_k (y_k \hat{y}_k)^2/(n - p - 1))$, which has an $F(p, n - p - 1)$ distribution. Shanmugam

(1987) [200] showed that $nc(s_2/k - 1)$ for a Pareto distribution is an $F(2, 2nc)$ where $s_2 = \min(x_2, x_3, \ldots, x_n)$.

A sum of n independent but not identically distributed (INID) F distributions occurs in wireless communications. For example, it is used as a tractable fading model to describe the combined effects of shadowing and multipath fading in wireless communication (Yoo et al. (2017) [235], Badarne et al. (2018) [10]). The instantaneous signal-to-noise ratio and the average bit error rate of coherent binary modulation schemes can also be modeled using INID F distribution.

16.7.1   *F*-TEST FOR EQUALITY OF VARIANCE

Suppose samples are drawn from two IID normal populations $N(\mu, \sigma_k^2)$ for $k = 1, 2$ with unknown variances. Our null hypothesis is $\sigma_1^2 = \sigma_2^2$. The test statistic for testing equality of variances is based on the central F distribution. The ratio of larger sample variance to the smaller one as $F = n_1 s_1^2/(n_2 s_2^2)$ has an $F(n_1 - 1, n_2 - 1)$ distribution, so that only the right tail is of interest (the F-score is greater than one).

16.7.2   *F*-TEST FOR EQUALITY OF REGRESSION

The T test is used to test the significance of a single regression coefficient in MLR, whereas an F test is used to test whether $\beta_1 = \beta_2 = \cdots = \beta_k = 0$ simultaneously. The F-test for equality of two sets of regression coefficients use the test statistic $[(SSE_{m+n} - SSE_n - SSE_m)/(k + 1)]/[(SSE_n + SSE_m)/(m + n - 2(k + 1))]$ where SSE_m, SSE_n are the error sum of squares from the first and second datasets of sizes m and n, SSE_{m+n} is the error sum of squares from the pooled data, and k is the number of regressors. It is assumed that both m and n are at least as large as k, preferably much larger (i.e., $m > k + 1$ and $n > k + 1$). This has an $F(k + 1, (m + n - 2(k + 1)))$ distribution.

16.8   SUMMARY

This chapter introduced the F distribution and its extensions. It finds extensive applications in ANOVA and related procedures. Special F distributions are briefly introduced. The chapter ends with a list of applications of F distribution in statistical inference.

CHAPTER 17

Weibull Distribution

17.1 INTRODUCTION

This distribution is named after the Swedish physicist Waloddi Weibull (1887–1979),[1] who invented it in 1937 in connection with strength of materials, although it was known to Fréchet (1927) and Rosin and Rammler (1933) [187]. The one-parameter Weibull distribution has PDF

$$f(x, a) = ax^{a-1}e^{-x^a}, \ x > 0. \tag{17.1}$$

It finds applications in reliability theory and survival analysis, quality control, strength of materials, ([94], [159]), voltage exceedances in electrical engineering, breakdown of electrical insulation, materials science,[2] etc. Although the Gaussian law is the most popular to describe the errors in prediction and forecasting models, the Weibull distribution is sometimes used when the error distribution is skewed. Weibull autoregressive conditional duration (WACD) model is one such model that is used in econometrics and finance. It is used in the design of wind turbines, model wind speed distributions, wind power analysis ([18], [221],[224]), and in fading channels in wireless communications. It can be used to describe the size of particles in motion (like raindrops), or those being grind, milled, crushed, or subjected to external pressure in metallurgy (for which another choice is the lognormal law).

One reason for the popularity of this distribution is that it works well in life testing applications even with small samples. Another reason is that the density function changes in shape drastically for various values of the shape parameter a. It is ideal in those situations where the data assumes a variety of distributional shapes, so that it can be used as a flexible model in situations with decreasing failure rate (DFR), constant failure rate (CFR), or increasing failure rate (IFR). The two-parameter Weibull distribution is obtained from (17.1) by a simple transformation $x = y/b$ as

$$f(x, a, b) = (a/b)(x/b)^{a-1}e^{-(x/b)^a}, \quad x > 0, \tag{17.2}$$

where b is the scale and a is the shape parameter.[3] Some fields use $1/b$ instead of b that results in the PDF

$$f(x, a, b) = ab^a(x)^{a-1}e^{-(bx)^a}. \tag{17.3}$$

---

[1]He described it in detail in 1951 with seven examples chosen from different fields like strength of steel, height of adult males, etc. The U.S. Air Force funded several of his works on this distribution until 1975.

[2]The shape parameter is known as Weibull modulus in materials science.

[3]k is used in place of a and λ is used in place of b in many fields like metallurgy, ecology, and reliability engineering.

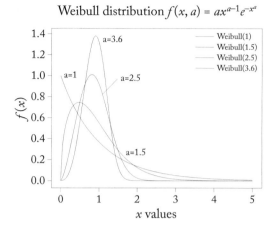

Figure 17.1: Weibull distributions.

The translated (three-parameter) Weibull distribution is obtained from the above using change of origin transformation as

$$f(x,a,b,c) = \frac{a}{b}((x-c)/b)^{a-1}e^{-((x-c)/b)^a}, \quad x \geq c, \tag{17.4}$$

where c is the location parameter, b is the scale parameter, a is the shape parameter. This reduces to the above form when $c = 0$. We denote it as WEIB (a,b,c) for location parameter $c \neq 0$, and WEIB(a,b) for $c = 0$.

17.2   RELATION TO OTHER DISTRIBUTIONS

It is easy to see that if $[(X-c)/b]^a$ has an exponential distribution, then X has a general Weibull distribution. The exponential distribution is a special case of the Weibull family when the shape parameter $a = 1$, and the two-parameter Rayleigh distribution when $a = 2$. It is positively skewed for small values of a, but resembles the Gaussian law for $a \approx 3.6$ because the skewness reduces to zero (see Figure 17.1). In most applications, the values of a are in the interval $(.5, 3.5)$. It is also related to the uniform distribution $U(0,1)$ as follows: if $X \sim U(0,1)$ then $Y = b(-\ln(X))^{1/a} \sim$ WEIB (a,b), and $Y = c + b(-\ln(X))^{1/a} \sim$ WEIB (a,b,c). It is a special case of the Frechet extreme value distribution with PDF

$$f(y,a,b) = \frac{a}{b}(y/b)^{-a-1}e^{-(y/b)^{-a}}. \tag{17.5}$$

If $X \sim$ WEIB $(a, 1/2)$, then $Y = \sqrt{X} \sim$ EXP $(1/\sqrt{a})$. If $X \sim$ WEIB (a,b) then $\log(X)$ has extreme value distribution with location parameter $\log(b)$ and scale parameter $1/a$. Similarly, if $X \sim$ WEIB (a,b,c), then $Y = -a\ln((X-c)/b)$ has general extreme value distribution. If X $\sim$

WEIB (a, b) then $Y = X^c$ belongs to the same family. As $a \to \infty$, this distribution converges to a Dirac delta function centered at $x = b$.

Discrete Weibull distribution (DWD) was obtained by Nakagawa and Osaki (1975) [120] with PMF

$$f(x; a, b, q) = q^{k^a} - q^{(k+1)^a}, \quad k = 0, 1, 2, \ldots; \quad q = \exp((-1/b)^a), \quad 0 < q < 1. \quad (17.6)$$

The minimum of n IID DWD random variables is identically distributed [3].

17.3 PROPERTIES OF WEIBULL DISTRIBUTION

It becomes the exponential distribution for shape parameter $a = 1$. Mode is $b(\frac{a-1}{a})^{1/a}$ for $a > 1$, and median is $b(\log 2)^{1/a}$. The mode tends to zero when $a \to 1$, in which case it tends to the exponential distribution. This tends to the limit 1 as $a \to \infty$. The three-parameter Weibull model has CDF

$$F(x; a, b, c) = 1 - \exp(-((x - c)/b)^a) \text{ for } x \geq c. \quad (17.7)$$

The SF $S(x; a, b, c) = \exp(-((x - c)/b)^a)$ is sometimes called the "stretched exponential function." Take $\log$ to get $\log(S(x; a, b, c)) = -((x - c)/b)^a$. Negate both sides and take $\log$ again to get $\log(-\log(S(x; a, b, c))) = a \log((x - c)/b) = a[\log(x - c) - \log(b)]$. Put $\log((x - c) = x'$, and $\log(-\log(S(x; a, b, c))) = y$ to get the form $y = mx + c$, which is the equation of a straight line. The inverse CDF is $F^{-1}(p; a, b, c) = c + b(-\ln(1 - p))^{1/a}$.

Problem 17.1   If $X_1, X_2, \ldots, X_n$ are IID Weibull random variables, find the distribution of $Z = \min(X_1, X_2, \ldots, X_n)$.

Problem 17.2   Prove that the hazard function of Weibull distribution is $h(t) = at^{a-1}/b^a$.

17.3.1 MOMENTS

The kth moment is given by

$$\mu'_k = E(x^k) = \int_0^\infty x^k f(x) dx = c + b^k \Gamma(k/a + 1). \quad (17.8)$$

From this we get the mean as $\mu' = \mu = c + b\Gamma(\frac{1}{a} + 1)$. When $1/a$ is an integer (say n), this becomes $\mu = c + n!b$. A similar reduction is possible when $1/a$ is a half-integer because $\Gamma(1/2) = \sqrt{\pi}$. The second moment is $b^2\Gamma(\frac{2}{a} + 1)$, from which the variance is

$$\sigma^2 = b^2 \left[\Gamma(2/a + 1) - \Gamma(1/a + 1)^2 \right]. \quad (17.9)$$

Using $\Gamma(n + 1) = n\Gamma(n)$, this becomes $(b^2/a)[2\Gamma(2/a) - \Gamma(1/a)^2/a]$. The analogous results for the alternate parametrization $b' = 1/b$ are obtained as $\mu = c + (1/b)\Gamma(\frac{1}{a} + 1)$ and $\sigma^2 = (1/b^2)\left[\Gamma(2/a + 1) - \Gamma(1/a + 1)^2\right]$.

The CV of WEIB (a, b) is $CV^2 = [\Gamma(2/a + 1)/\Gamma^2(1/a + 1)] - 1 = (2\Gamma(2/a)/[\Gamma(1/a)$ $\Gamma(1/a + 1)]) - 1$ (Table 17.1). The MGF does not have closed form except for special values of the shape parameter due to the presence of the exponent term. However, $E[\log(x)]$ has closed form as $E[\log(x)] = b^t \Gamma(t/a + 1)$. See Lai (2014) [124] for properties of generalized Weibull distributions.

Problem 17.3   Find $\ln(-\ln(\Pr(X > x))$ for a Weibull distribution.

Problem 17.4   Find the hazard function $h(t) = f(t)/S(t)$ for the Weibull distribution.

17.3.2   RANDOM NUMBERS

As $F(x) = 1 - e^{-(x/b)^a}$, we could generate random numbers using uniform pseudorandom numbers in $[0, 1]$ as $u = e^{-(x/b)^a}$ (as U and 1-U are identically distributed), which on rearrangement becomes $x = b(-\log(u))^{1/a}$. Notice that the log function takes negative values for the argument in $[0, 1]$. Hence, the $-\log(u)$ maps it into the positive interval.

17.3.3   FITTING

The logarithm of the PDF simplifies considerably. This technique is used in estimating the unknown parameters. For example, Menon (1963) [152] obtained an estimate of a as the positive square root of

$$\hat{a}^2 = \pi^2/[6 \sum_{k=1}^{n} (\log(x_k) - \log(\bar{x}))^2/(n-1)]. \tag{17.10}$$

This estimate is unbiased and approaches asymptotic efficiency rapidly. Higher-efficiency estimators have been proposed by several researchers (Kübler (1979) [122], Kappenman (1985) [113]). As the range of the distribution is $x > 0$, care must be taken to ensure that none of the data values are too close to zero. Such values can be discarded using a cutoff limit to prevent memory overflow problems. Weibull proposed parameter estimates using the smallest sample value $x_{(1)}$, finding the quantities $R_j = \sum_{k=1}^{n}(x_k - x_{(1)})^j/n$ for $j = 1, 2$, and solving $f(a^*) = R_2/R_1^2$ (which he tabulated). Using the expected value of the first order statistic $x_{(1)}$, along with first and second moments, Cohen et al. (1984) [50] proposed a modified MoM estimation technique. Fractional moments can also be used to estimate the parameters (Mukherjee and Sasmal (1984) [158]).

A Weibull plot can reveal if the data came from this distribution. For this, plot $\ln(x)$ along the X-axis and $\ln(-\ln(1 - F(x)))$ along the Y-axis, where F is the empirical CDF obtained from a random sample. A straight line indicates Weibull parent population.

Example 17.5   MD of Weibull distribution   Find the mean deviation of Weibull distribution.

Solution: We know the CDF is $1 - \exp(-(\frac{x-c}{b})^a)$. This gives

$$MD = 2 \int_c^m [1 - \exp(-(\frac{x-c}{b})^a)]dx, \tag{17.11}$$

Table 17.1: Properties of Weibull distribution

Property	Expression	Comments
Range of X	$0 \leq x < \infty$	Continuous
Mean	$\mu = b\Gamma(1 + 1/a)$	
Mode	$b(1 - 1/a)^{1/a}, a > 1$	$\rightarrow b$ as $a \rightarrow \infty$
Median	$b(\log 2)^{1/a}$	
Variance	$\sigma^2 = b^2[\Gamma(1 + 2/a) - \Gamma(1 + 1/a)^2]$	$= b^2\Gamma(1 + 2/a) - \mu^2$
CV	$(\Gamma(1 + 2/a)/\Gamma(1 + 1/a)^2 - 1)^{1/2}$	$\rightarrow 0$ as $a \rightarrow$ large
Skewness	$\gamma_1 = (\Gamma(1 + 3/a)b^3 - 3\mu\sigma^2 - \mu^3)/\sigma^3$	
Kurtosis β_2	$3c(c + 1)(a + 1)(2b - a)/[ab(c + 2)(c + 3)]$	$c = a + b$
Moments	$\mu'_r = b^r\Gamma(1 + r/a)$	
MD	$(2/a)\gamma(m^a; 1/b^a, 1/a)$	$m = c + b\Gamma(1 + 1/a)$
MGF	$\sum_{k=0}^{\infty}(bt)^k\Gamma(1 + k/a)/k!$	
ChF	$\sum_{k=0}^{\infty}(bit)^k\Gamma(1 + k/a)/k!$	
Additivity	$\sum_{i=1}^{m} Y(a, b_i) = Y(a, \prod_{i=1}^{m}b_i)$	IID Y = log(WEIB)
Recurrence	$(1 + 1/a)(x/b) \exp(-(x/b)^a[1 - x/b])$	$f(x; a + 1, b)/f(x; a, b)$
Tail area	$\exp(-(x/b)^a)$	

The mode of $WEIB(a, b) \rightarrow b$ as a becomes large ($0.9974386\,b$ for $a = 20, 0.999596\,b$ for $a = 50$), but median $\rightarrow b$ much slower ($0.981841\,b$ for $a = 20, 0.992697\,b$ for $a = 50$, $0.9995964\,b$ for $a = 908$). The MFG and ChF of logarithm of the Weibull variate are more tractable, $E[\exp(ln(X)t)] = b^t\Gamma(1 + t/a)$.

where $m = c + b\Gamma(1 + \frac{1}{a})$ is the mean. Split this into two integrals, and integrate the first term to get $2b\Gamma(1 + \frac{1}{a})$. The second integral is $-2\int_c^m \exp(-(\frac{x-c}{b})^a)dx$. This can be converted into an incomplete gamma function by a simple substitution $y = ((x - c)/b)^a$, so that $dx = (b/a)y^{1/a-1}dy$, to get $2b/a\gamma(\Gamma(1 + 1/a)^a, 1; 1/a)$ where $m = c + b\Gamma(1 + 1/a)$. Now combine with the first term to get the MD as $2[K + \gamma(K^a, 1/b^a, 1/a]$, $K = b\Gamma(1/a + 1)$.

Example 17.6 Functions of IID Weibull distributions   If X, Y are IID Weibull $(2, b)$ with PDF $f(x, b) = \frac{2}{b^2}xe^{-x^2/b^2}$, for $x \geq 0$, show that $Z = XY/(X^2 + Y^2)$ has PDF $f(z) = 2z/\sqrt{1 - 4z^2}$, which is independent of b.

Solution: Put $x = r\cos(\theta)$, and $y = r\sin(\theta)$, so that $x^2 + y^2 = r^2$, and $\theta = \tan^{-1}(y/x)$. The Jacobian of the transformation is r. Hence, the joint PDF of r and θ is $f(r, \theta) = \frac{4r^3}{b^4}\sin(\theta)\cos(\theta)e^{-r^2/b^2}$. Using $2\sin(\theta)\cos(\theta) = \sin(2\theta)$ this becomes $f(r, \theta) = \frac{2r^3}{b^4}\sin(2\theta)e^{-r^2/b^2}$. The PDF of θ is obtained by integrating out r as

$$f(\theta) = \frac{2\sin(2\theta)}{b^4} \int_0^\infty r^3 e^{-r^2/b^2}\, dr. \tag{17.12}$$

Put $r^2/b^2 = t$, so that $r\, dr = (b^2/2)dt$. Then $f(\theta) = \frac{2\sin(2\theta)}{b^4}\frac{b^4}{2}\int_0^\infty te^{-t}dt = \sin(2\theta)$. Putting the values of x and y in Z gives $Z = \frac{1}{2}\sin(2\theta)$ so that $\frac{\partial z}{\partial \theta} = \cos(2\theta) = \sqrt{1 - 4z^2}$. Hence, $f(z) = 2z/\sqrt{1 - 4z^2}$, $-\frac{1}{2} \le z \le \frac{1}{2}$.

17.4   GENERALIZED WEIBULL DISTRIBUTION

Weibull distribution can be generalized either by adding one or more additional parameters (conveniently to the CDF or SF) to the baseline distribution, or by other techniques like size-biasing, truncation, exponentiation, etc. The inverse Weibull distribution (IWD), which is the distribution of $Y = 1/X$, finds applications in biological sciences, reliability, geology, and metallurgy. The gradual degradation and eventual failure under load is a phenomena observed in several engineering industries. This is extensively used in mechanical engineering, metallurgy, and transportation. Metal fatigue is usually divided into mechanical and metallurgical faults (Pook (2007) [175]. Metallurgical faults are major cause for concern as it could result in catastrophic machine failures. These are modeled using extended Weibull models. The characteristics of a component, a part or a product like the infant mortality, lifetime and wear-out periods can be used to determine the cost effectiveness, and find maintenance schedules and costs using the IWD. The CDF is given by

$$G(x; a, b) = \exp(-(a/x)^b), \quad x > 0, \ a, b > 0. \tag{17.13}$$

Differentiate wrt x to get the PDF as

$$g(x; a, b) = ba^b/x^{b+1}\exp(-(a/x)^b), \quad x > 0, \ a, b > 0. \tag{17.14}$$

The inverse of CDF is easily obtained as $x = (-a^b/\log(p))^{1/b}$. The inverse Weibull-geometric distribution (IWGD) is an extension with CDF

$$F(x; a, b, p) = (1 - \exp(-(bx)^a))/(1 - p\exp(-(bx)^a))), \tag{17.15}$$

where $p \in (0, 1)$. Zacks (1984) [237] considered a system as having a "two-phase life" in which the first phase has a CFR until a change point time t, followed by the second (wear-out) phase with a DFR using an exponential Weibull distribution.

Right-truncated Weibull distributions (RTWD) find applications in epidemiology in modeling the progression of a disease like COVID-19. Moments of RTWD can be found in McEwen and Parreson (1991) [149], and parameter estimates in Mittal and Dahiya (1989) [154]. Mud-holkar and Srivastava (1993) [157] proposed a three-parameter Weibull distribution with CDF $F(x; a, \lambda, b) = (1 - \exp(-\lambda x^a))^b$ where a, b are the shape parameters, and λ is the scale parameter. This coincides with two-parameter Weibull distribution for $b = 1$. See Elgarhy et al. (2017) [62] for an exponentiated Weibull-Exponential distribution, Barreto et al. [19] for Weibull-geometric distribution, Khan and Jan (2016) [115] for inverse Weibull-geometric distribution, Drapella (1993) [60] for a complementary Weibull distribution, Nassar et al. (2018) [165] for an extension of Weibull distribution, and Murthy et al. (2003) [159] and Lai (2014) [124] for a detailed discussion of generalized Weibull distributions.

17.5 FITTING

Parameters can be estimated using numerical methods, analytical methods (MLE, method of moments, etc.), or graphical methods (empirical CDF plot, Weibull probability plot, empirical hazard function plot). The MLE of a takes the form $\sum_{k=1}^{n} x_k^a \ln(x_k) / \sum_{k=1}^{n} x_k^a - 1/a - (1/n) \sum_{k=1}^{n} \ln(x_k) = 0$. This shows that the estimate of the shape parameter involves raw moments of order k, and iterative methods have to be used for estimation. The MLE of b given a is $\hat{b} = \frac{1}{n} \sum_{k=1}^{n} x_k$. Balakrishnan and Kateri (2008) [12] used a graphical method that can also hint on the existence of MLE estimates. See Woo, Ali, and Nadarajah (2005) [233] for the distribution of the ratio $X/(X + Y)$ for Weibull random variables, and Nadarajah and Kotz (2006) [162] for the product and ratio of Weibull random variables.

17.6 APPLICATIONS

The Weibull distribution is the most widely used probability distribution in reliability engineering. An ideal situation where this distribution finds applications is in the failure of an item or component that is subjected to "wear and tear" over continuous use. This is called *wear-out failure* in manufacturing engineering. Machines may fail due to natural causes, mechanical, electrical, or material problems, wrong handling or operations, design faults, or other reasons. It is also used to model the recall of machines (like automobiles) and parts (like batteries used in phones, e-vehicles, compressors used in air-conditioners, etc.) based on warranty time failure, recall of medicines due to harm or death to patients, maintenance planning for machinery or parts (optimal spare stocking), warranty analysis, batch defect analysis, and risk analysis.

In those situations where components or constituents form a "whole-part" relationship or are coupled together, it does not make a difference between applying the Weibull model to the whole or to the parts. Weibull (1961) [227] lists several applications of this distribution, and is a good starting point for anyone interested in its applications.

17.6.1   WEIBULL PLOTS

Weibull probability plots (WPP) are visualization tools for product analysis that have failed over an arbitrary time-period. It involves plotting the data, interpreting the graphs, and forecasting failures. It is applicable to machines and parts that are used intermittently (like computers, vehicles, solar power plants) or continuously (transformers in power supply, nuclear power stations). It is used in mechanical engineering, automobile engineering, automated manufacturing systems, material science, polymer science, fiber optics, and nanotechnology. Time to failure should be measured logically for each product (in cycle-starts, hours of run-time, miles driven before breakdown, etc). A time-origin is then fixed, a scale for measuring the passage of time decided, and the exact meaning of failure (a cutoff may have to be defined in some situations like batteries, compressors, ball-bearings, conveyors, suspension springs in automobiles, etc. to decide whether it is totally unusable) defined. A log-log plot (called WPP) is then generated with log of time to failure (TTF) along the X-axis and log of the percentage of items that have failed along the Y-axis. Technically, plot $\ln(x)$ along the X-axis and $\ln(-\ln(1 - F(x)))$ along the Y-axis, so as to linearize the SF. This percentage can be calculated only if the age (in terms of time units) of both failed and still-working parts are reckoned. Then WPP can reveal the percentage failure rate (PFR) of items. The slope (m) of the best-fit line in a least-square or least-absolute sense is found. It is very informative as it could provide a clue to the failure mechanism. This describes the Weibull failure distribution. If $m > 1$, it is an indication of "wear-out" failure, whereas $m = 1$ means random failure.

17.6.2   GEOLOGY

Weibull distribution finds applications in modeling volcanic eruptions, time between occurrences of minor quakes or expected time delay in the occurrence of a quake of magnitude higher than a specified value in a locality in geology. Past data on earthquakes are used to fit the model. Hence, we get different Weibull models for different regions, as well as for different cutoff magnitude. As the Richter magnitude is a logarithmic scale, and the highest magnitude earthquake on land ever recorded is 8.2, the chosen value of m must be lower than this [40]. The seismic risk is an uncertainty measure used by geologists as

$$R(t; a_m, b_m) = 1 - \exp(-(t/b_m)^{a_m}) \tag{17.16}$$

which gives the probability of observing at least one quake of magnitude greater than m in time t.

17.6.3   RELIABILITY ENGINEERING

Static mechanical properties of metallic materials are modeled using Gaussian, skewed normal, log-normal or Weibull distributions. For example, the tensile strength, hardness and impact strength of metals can be modeled using Gaussian law while the 2-parameter Weibull distribution is preferred for fracture toughness of metals (like tensile property of aluminum alloys), and 3-parameter Weibull distribution is preferred for fatigue life of metallic materials (like ferrous alloys). The lifetimes of several components like bearings, capacitors, and dielectrics can be modeled by Weibull distribution as well. The failure rate in such systems is proportional to a power of time of continuous operation (wrt a fixed starting point). Several practical applications of this model involve a chain of interlocking links as in fiber optics, polymer science, photovoltaic grids, nanotechnology, to mention a few. The links can be linear (1D), spatial (2D), or volumetric (3D), and can have slightly different strengths. A Weibull model uses the length of the chain interpreted in an appropriate manner. The PDF then captures the probability of any link failing at or below an applied stress level, so that the CDF can be written as

$$F(s; a, s_0) = 1 - \exp(-s/s_0)^a, \tag{17.17}$$

where the scale parameter s_0 denotes the severity of flaws. Assuming that the system will fail if any of the components in a gauge of length L fails, we could model the survival probabilities using the "links in series" model. An implicit assumption is that each link fails independently of its neighboring links. This gives the survival probability CDF as

$$F(s; a, s_0) = 1 - \exp(-L(s/s_0)^a), \tag{17.18}$$

where L is the portion of the length subjected to the stress. Values of $a < 1$ indicates DFR over the passage of time because defective items are weeded out and only healthy ones are used subsequently. The case $a = 1$ indicates CFR over time (which is the exponential distribution case), and $a > 1$ indicates IFR. As aforementioned, $a = 2$ results in Rayleigh distribution, and $a \approx 3.6$ resembles a normal law. Thus, a wide variety of distributional shapes can be realized. The mean residual life function can be expressed in terms of incomplete gamma functions (Lai (2014) [124]).

17.7   SUMMARY

This chapter discussed the Weibull distribution, and its generalizations. Only a few among the large number of applications are discussed due to space constraints, but it is increasingly being used in medical sciences, data science, epidemiology, and other emerging fields. Interested readers can find many resources in the reference section.

CHAPTER 18

Rayleigh Distribution

18.1 INTRODUCTION

This distribution is named after the British physicist John Baron Rayleigh (1842–1919)[1] who introduced it in 1880 in connection with interference of random phase harmonic oscillations in a communication channel. It is also used in reliability engineering, radar and microwave systems, image recognition, wind energy modeling, design of electro-vacuum devices, wave heights modeling in oceanography, acoustics, magnetic resonance imaging (MRI), and nutrition (nutrient levels in fruits and vegetables) (Shanmugam (2020), [204]). It is a continuous distribution with support $[0, \infty)$, and PDF

$$f(x; a) = (x/a^2) \exp(-x^2/(2a^2)), \quad x \geq 0, \tag{18.1}$$

and belongs to the general exponential family. It can also be written as

$$f(x; a) = (1/a) \quad (x/a) \exp(-\frac{1}{2}(x/a)^2)), \quad x \geq 0, \tag{18.2}$$

which shows that a acts as a scale parameter. This form is used for extensions of this distribution. Put $c = 2a^2$ to get another form

$$f(x; c) = (2x/c) \exp(-x^2/c), \quad x \geq 0, \tag{18.3}$$

It is reparametrized as $a^2 = 2b^2/\pi$ in some fields, so that $E(X) = a$.

This distribution can be considered as the distribution of the radial distance of a point on the bivariate normal surface (with zero means) from the origin. In other words, it is the distribution of $\sqrt{X^2 + Y^2}$ where (X, Y) have a joint bivariate normal distribution centered at zero. The X and Y represent the distances from a target (which is assumed to be located at $(0, 0)$) that a projectile lands in warfare, and a probe lands in space missions. The χ-distribution with two DoF and Rayleigh distribution are exactly identical. An alternate form can be obtained by the linear transformation $y = (\sqrt{b}/[a\sqrt{2}])x$, so that $dx = a\sqrt{2}/\sqrt{b}dy$ to get the PDF as

$$f(y; b) = 2(y/b) \exp(-y^2/b), \quad y \geq 0. \tag{18.4}$$

It is called Wigner's distribution or Rayleigh–Wigner distribution in nuclear science. We denote it as R(a) or Rayleigh(a).

---

[1]He was also known as John William Strutt and Lord Rayleigh.

18.2   RELATION TO OTHER DISTRIBUTIONS

If X and Y are IID $N(0, \sigma^2)$ random variables, then $R = \sqrt{X^2 + Y^2}$ has Rayleigh(σ) distribution. More generally, if $X \sim N(\mu_x, \sigma_x^2)$ and $Y \sim N(\mu_y, \sigma_y^2)$ are IID normal random variables, $R = \sqrt{((X - \mu_x)/\sigma_x)^2 + ((Y - \mu_y)/\sigma_y)^2}$ has Rayleigh(1) distribution. When the means are non-zero, R has a Rice distribution with PDF

$$f(x; A, a) = (x/a^2) \exp(-\frac{1}{2}((x^2 + A^2)/a^2)) I_0(Ax/a^2), \quad x \geq 0, \qquad (18.5)$$

where $A = \sqrt{\mu_1^2 + \mu_2^2}$. It follows from this that the magnitude $|z|$ of a standard complex normally distributed variable with zero mean has a Rayleigh distribution. If $X \sim$ Rayleigh(1), then X^2 is χ_2^2, so that the χ distribution with DoF $= 2$ is equivalent to the Rayleigh(1). If $X \sim$ Rayleigh(a), then X^2 is EXP($2a^2$) with PDF $f(x; a) = (1/(2a^2)) \exp(-x/(2a^2))$. As χ_2^2 and gamma distributions are related, it follows that if $X_i \sim$ Rayleigh(b), then $\sum_i X_i^2$ is gamma distributed. It is related to $U(0, 1)$ as $X = a(-2\ln(1 - U))^{1/2}$. As U and $1 - U$ are identical, $X = a(-2\ln(U))^{1/2}$ also is Rayleigh distributed. Alternatively, if X has a standard Rayleigh distribution, then $U = 1 - \exp(-X^2/2)$ is U(0,1). If $X \sim$ EXP(λ), then $Y = \sqrt{X} \sim$ Rayleigh($1/\sqrt{2\lambda}$). The Chi-distribution and Weibull distributions are generalizations of the Rayleigh distribution, as also the Rice distribution when the mean is non-zero. When the shape parameter $a = 2$, and scale parameter $b = \sigma\sqrt{2}$ for a Weibull distribution, we get a Rayleigh(σ) distribution. As the Weibull and gamma distributions are related, this distribution can be considered as a special case of gamma law. Thus, Rayleigh distribution provides an alternate choice to model lifetime of devices that have a linearly increasing failure rate. Put $b = 1/(2a^2)$ to get

$$f(x; b) = 2bx \quad \exp(-bx^2), \quad x \geq 0, \qquad (18.6)$$

which has mode $1/\sqrt{2b}$ with modal value $\exp(-1/2)\sqrt{2b} = 0.85776388\sqrt{b}$. This is the distribution of 1-D energy level spacings of nuclear spectra at high energies. It is known as Wigner's surmise distribution in high-energy physics[2] (Hillery et al. (1984) [92], Wojnar (2012) [231]).

Problem 18.1   If X_k are IID Rayleigh(a), find the distribution of $Y = \sum_{k=1}^{n} X_k^2$.

Problem 18.2   If $X \sim$ EXP(λ), find the distribution of $Y = \sqrt{2\lambda X}$.

18.3   PROPERTIES OF RAYLEIGH DISTRIBUTION

The standard Rayleigh distribution (SRD) is obtained by putting $a = 1$ with PDF $f(x; 1) = x \exp(-x^2/2)$. The PDF first increases and then decreases with mode at $x = a$ with modal value $(a/a^2) \exp(-a^2/(2a^2)) = \exp(-0.5)/a = 0.60653066/a$. The PDF is first concave downward

---

[2]E. P. Wigner surmised that the distribution of widths and spacings of nuclear resonance levels is the Gaussian law multiplied by $x > 0$ (which statisticians call "size-biased half-Gaussian law"), making small distances less likely.

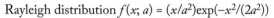

Rayleigh distribution $f(x; a) = (x/a^2)\exp(-x^2/(2a^2))$

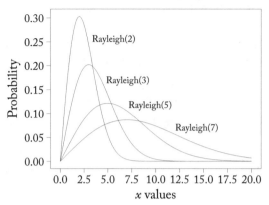

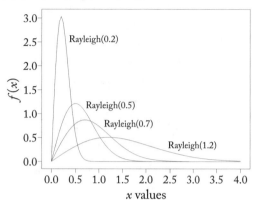

Figure 18.1: Rayleigh distributions.

and then upward with point of inflection at $x = a\sqrt{3}$ (see Figure 18.1). The affine transform $cx + d$ yields the same distribution. Archer (1967) [5] obtained the distribution of the product of two independent Rayleigh random variables, which was extended to an arbitrary number of random variables by Salo, El-Sallabi, and Vainikainen (2006) [189].

Problem 18.3   If $X \sim$ Rayleigh(a), prove that the distribution of $Y = X^2$ is exponential EXP($1/(2a^2)$).

18.3.1   MOMENTS AND GENERATING FUNCTIONS

Ordinary moments can be obtained in terms of gamma function ($n = 2$, σ=a) as E[X^k] =

$$E(X^k) = \mu'_k = (1/a^2) \int_0^\infty x^{k+1} \exp(-x^2/(2a^2))dx. \tag{18.7}$$

Make the substitution $y = x^2$ so that $dy = 2xdx$. The range of integration remains the same, and we get $E(X^k) =$

$$(1/a^2) \int_0^\infty y^{(k+1)/2} \exp(-y/(2a^2))dy/(2\sqrt{y})$$

$$= (1/(2a^2)) \int_0^\infty y^{(k/2+1)-1} \exp(-y/(2a^2))dy. \tag{18.8}$$

The last integral is of the form $\int x^{p-1} \exp(-mx)dx$ so that $E(X^k) = (1/(2a^2))\Gamma(k/2 + 1)(2a^2)^{k/2+1}$, where $\Gamma()$ is the complete gamma function. Cancel out $2a^2$ to get

$$\mu'_k = 2^{k/2}\Gamma(k/2 + 1)a^k = (2a^2)^{k/2}\Gamma(k/2 + 1) = ka^k\Gamma(k/2)2^{k/2-1}. \tag{18.9}$$

Put $k = 1$ and use $\Gamma(3/2) = \sqrt{\pi}/2$ to get the mean as $\mu = a\sqrt{\pi/2} \approx 1.25331a$. Next, put $k = 2$ to get $\mu'_2 = 2\Gamma(2)a^2 = 2a^2$, from which the variance $\sigma^2 = (4 - \pi)a^2/2 \approx 0.4292036a^2$. It follows that the standard deviation is $\sigma \approx 0.6551364a$. This shows that $\sigma^2 < \mu^2$ or equivalently $\sigma/\mu \approx 0.5227232$. The mode is a, and the median is $a\sqrt{\ln 4} = a\sqrt{2\ln 2} \approx 1.17741a$. The CV is $\sqrt{4/\pi - 1} \approx 0.5227232$, and skewness coefficient is $2(\pi - 3)\sqrt{\pi}/(4 - \pi)^{3/2} \approx 0.631111$, which is independent of the parameter a. As the skewness is ≈ 0.63111066, it is always positively skewed. The kurtosis coefficient is $(32 - 3\pi^2)/(4 - \pi)^2 \approx 3.24509$, showing that the excess kurtosis is .245.

Problem 18.4   Prove that the third raw moment is $\mu'_3 = 3a^3\sqrt{\pi/2}$ and central moment is $\mu_3 = a^3\sqrt{\pi/2}(\pi - 3) \approx 0.17746a^3$.

Problem 18.5   Prove that the fourth moments are $\mu'_4 = 8a^4$ and $\mu_4 = a^4(32 - 3\pi^2)/4 \approx 0.597796699a^4$.

The CDF is $1 - \exp(-x^2/2a^2) = 1 - a\sqrt{2\pi}\phi(x/a)$, from which the IDF follows as $F^{-1}(p; a) = a(-2\log(1 - p))^{1/2}$. The MGF is

$$M_x(t) = 1 + at\exp(a^2t^2/2)\sqrt{\pi/2}[\text{erfc}(at/\sqrt{2}) + 1]. \tag{18.10}$$

The quartiles are $Q_1 = a\sqrt{4\ln(2) - 2\ln(3)} \approx 0.7585276a$, $Q_2 = a\sqrt{2\ln(2)} = 1.17741a$, $Q_3 = a\sqrt{4\ln(2)} \approx 1.66511a$. In general if $X_1, X_2, \ldots, X_n$ are independent normal random variables $N(0, \sigma^2)$, the distribution of $X = (X_1^2 + \cdots + X_n^2)^{1/2}$ is given by

$$f(x; n, \sigma) = \frac{2}{(2\sigma^2)^{n/2}\Gamma(n/2)} x^{n-1} e^{-x^2/2\sigma^2} \text{ for } x > 0. \tag{18.11}$$

See Table 18.1 for further properties.

Problem 18.6   If $f(x) = (x/\sigma^2)\exp(-x^2/(2\sigma^2))$, $\quad 0 < x < \infty$ prove that the hazard function is x/σ^2.

Example 18.7   MD of Rayleigh distribution   Find the MD of the Rayleigh distribution.

Solution: As the CDF is $1 - \exp(-x^2/2a^2)$, the MD is given by

$$\text{MD} = 2\int_0^m [1 - \exp(-x^2/2a^2)]dx \text{ where } m = a\sqrt{\pi/2}. \tag{18.12}$$

Split the integral into two parts. The first one integrates to $2m$. The second one is $-2\int_0^m \exp(-x^2/2a^2)dx$. Put $y = x^2/(2a^2)$ so that $dy = x/a^2 dx$. The upper limit of integration becomes $m^2/(2a^2) = \pi/4$. We get

$$\text{MD} = 2m - 2\int_0^{\pi/4} y^{\frac{1}{2}-1}e^{-y}dy = 2[m - P(\pi/4, 1/2)], \tag{18.13}$$

where $P()$ is the incomplete gamma integral. Put the value of m to get the MD.

Problem 18.8   If $X \sim \text{Rayleigh}(a)$ prove that the distribution of $Y = X^2$ is $\text{EXP}(2/(2a^2))$.

Table 18.1: Properties of Rayleigh distribution $(x/a^2)\,e^{-x^2/(2a^2)}$

Property	Expression	Comments
Range of X	$0 \leq x < \infty$	Continuous
Mean	$\mu = a\sqrt{\pi/2}$	$\approx 1.253314\,a$
Mode	a	
Median	$a\sqrt{\ln(4)}$	$\approx 1.17741\,a$
Variance	$\sigma^2 = (2 - \pi/2)a^2 = 2a^2 - \mu^2 = \mu^2(4/\pi - 1)$	$\approx 0.27324\mu^2$
Skewness	$\gamma_1 = 2(\pi - 3)\sqrt{\pi}/(4 - \pi)^{3/2}$	≈ 0.63111
Kurtosis	$\beta_2 = (32 - 3\pi^2)/(4 - \pi)^2$	
MD	$2[a\sqrt{\pi/2} - \gamma(\pi/4, 1/2)]$	$\gamma = $ incomplete Γ
Quartiles	$Q_1 \approx 0.75853\,a$	$Q_3 \approx 1.66551\,a$
CV	$\sqrt{4/\pi - 1}$	≈ 0.522723
Moments	$\mu_r' = 2^{r/2}a^r\Gamma(r/2 + 1)$	
MGF	$1 + at \exp(a^2t^2/2)\sqrt{\pi/2}[\text{erfc}(at/\sqrt{2}) + 1]$	
ChF	$1 - ait \exp(-a^2t^2/2)\sqrt{\pi/2}[\text{erfc}(at/\sqrt{2}) - i]$	
Tail area	$\Pr[X > x] = \exp(-x^2/2a^2)$	

The ratio $\sigma/\mu = 0.5227232$ shows that this distribution is under-dispersed.
The mean-median-mode inequality is mode < median < mean.

18.4   FITTING

Let $x_1, x_2, \ldots, x_n$ be a random sample of size $n > 1$ from the Rayleigh distribution. The likelihood function is

$$L(x_1, x_2, \cdots, x_n; a) = (a)^{-2n} \prod_{k=1}^{n} (x_k) \exp([-1/(2a^2)] \sum_{k=1}^{n} x_k^2). \qquad (18.14)$$

The MLE of a is $\hat{a} = [(\sum_{k=1}^{n} x_k^2)/(2n)]^{1/2}$, which is biased. In other words, the square of the MLE estimate is half the quadratic mean of the sample. It can be made unbiased by multiplying it with $K = \Gamma(n)\sqrt{n}/\Gamma(n + 0.5)$. As the quadratic transformation $Y = X^2$ results in an exponential distribution, those methods discussed in Chapter 3 of Part I may be used to get confidence intervals.

18.5   GENERALIZED RAYLEIGH DISTRIBUTION

The variate can be scaled to get the two-parameter Rayleigh distribution with PDF

$$f(x; a, b) = (a/b)(x/b)^{a-1} \exp(-(x/b)^a) \qquad (18.15)$$

which reduces to (18.2) for $a = 2$, and $b = \sigma\sqrt{2}$. If $X \sim N(\mu_x, \sigma^2)$ and $Y \sim N(\mu_y, \sigma^2)$ are IID normal random variables, $R = \sqrt{X^2 + Y^2}$ has a Rice distribution with PDF

$$f(x; A, a) = (x/a^2) \exp(-\frac{1}{2}((x^2 + A^2)/a^2))I_0(Ax/a^2), \quad x \geq 0, \tag{18.16}$$

where $A = \sqrt{\mu_1^2 + \mu_2^2}$. This is also the distribution of $|X|$ in the complex case. The moments are given by $\mu_r = a^r 2^{r/2}\Gamma(r/2 + 1)L_{r/2}(-A^2/(2a^2))$ where $L_n(x)$ is the Laguerre polynomial.

The inverse Rayleigh distribution (IRD), which is the distribution of $Y = 1/X$, finds applications in reliability, tensile strength of materials and composites (carbon fibers, semiconductor wafers, etc.), and quality control. This has CDF $G(x; a) = \exp(-a/x^2)$ for $a, x > 0$. Leão et al. (2013) [133] introduced a two-parameter beta inverse Rayleigh distribution (BIRD) with CDF

$$G(x; a, b) = (1/B(a, b)) \int_0^{\exp(-a/x^2)} t^{a-1}(1 - t)^{b-1}dt, \tag{18.17}$$

which has closed-form expression $G(x; a) = (2/\pi)\arcsin(-a/(2x^2))$ when $a = b$. The slashed Rayleigh distribution (Iriarte et al. (2015) [97]) is defined as $Y = X/U^{1/k}$ where X is a one-parameter Rayleigh(a) variate, U is an independent standard uniform variate, and k is a positive real number. The PDF is

$$f(y; a, k) = (k(2a)^{k/2}/y^{k+1})\Gamma(k/2 + 1)F(y^2/(2a); k/2 + 1, 1)] \tag{18.18}$$

$$= k^2(2a)^{k/2}/[2\,y^{k+1}]\Gamma(k/2)F(y^2/(2a); k/2 + 1, 1), \quad y > 0, \tag{18.19}$$

where $F(x; a, b)$ denotes the CDF of a gamma distribution with shape parameters a and b. This has raw moments $\mu_r = (2a)^{r/2}\Gamma(r/2 + 1)k/(k - r)$, from which the mean is $\mu = k/(k - 1)\sqrt{a\pi/2}$ for $k > 1$ and $\mu_2' = k/(k - 2)\,2a$ for $k > 2$. Size-biased Rayleigh distribution has PDF

$$f(x; a) = (x^2/(a^3\sqrt{\pi/2}))\exp(-x^2/(2a^2)), \quad x \geq 0, \tag{18.20}$$

which is the Maxwell distribution (Chapter 19). Size-biasing using $E(1 + kx)$ (Chapter 1 of Part I; p. 6) results in a mixture of Rayleigh and Maxwell distributions. Weighted distributions could also be developed using moments, fractional moments and inverse moments. Stability of the Rayleigh distribution is discussed in [220], and a modified inverse Rayleigh distribution (MIRD) in Khan (2014) [116].

18.6   APPLICATIONS

Some high-end digital scanners capture images in complex form (with real and imaginary parts). As most image visualization software use real data, these are converted into magnitude images using $|z| = \sqrt{x^2 + y^2}$ where x and y are the real and imaginary parts, so that z has a Rayleigh distribution, if the components are approximately Gaussian. A circularly symmetric complex Gaussian distribution is also assumed in some of the data transmission applications. If the variance is assumed to be one in both the real and imaginary parts, the absolute value has a Rayleigh distribution with a variance 2. This can be normalized by dividing by 2 so that the real and imaginary part will have a variance of 1/2 each.

Rayleigh distribution is used for speckle or tissue image estimation in ultrasound and MRI [55], [192]. A mixture of Rayleigh distribution is used to detect targets in radar images [163], especially synthetic aperture radars. Some radar software use sliding-window based constant false alarm rate (CFAR) technique to detect targets. Rayleigh distribution is also a popular choice for statistical loads of structural components subjected to random stresses, and to approximate the spectral distribution of black body radiation (for which the Maxwell distribution is a better choice). Janssen (2014) [99, 100] used the Rayleigh(1/2) distribution to model freak wave warning systems used in weather prediction, and envelope wave heights in time series modeling. Applications in economics and social sciences can be found in Wojnar (2012) [231], who also discusses its utility in public transport of some countries and biomass pyrolysis in [58].

18.6.1   RAYLEIGH FADING

It is the name given to the form of signal fading in a communication channel. Some signals like radio waves (used in wireless cellular phones, GPS, and some navigation systems) have reflectivity property when they hit the boundary or surface of an object or dense medium like water.[3] The signals fade according to the Rayleigh law when there is a large number of oblique reflections present in a channel. Rayleigh fading model (RFM) uses a statistical approach to evaluate the propagation and reception of signals using the absolute value of the amplitude. While the minimum distance path is the direct (line of sight (LOS)) path, radio signals propagated through terrestrial environment will inevitably undergo reflections and refractions resulting in multiple signals at the receiver-end.[4] The overall received signal is a combination of all the signals that have reached through different paths. Because the radio signals travel nearly at the speed of light, the time delay between the receipt of the earliest and latest signal may be of the order of nanoseconds. Thus, the parameters of the Rayleigh distribution to be chosen depends on the distance between the transmitter and receiver. Signal processing applications use the phase of the received signal to identify actual (primary or dominant) signal from noise. Signals that are all "in phase" with each other will add together to the overall signal, and those "out of phase" will subtract. The RFM works best when there is no dominant signal (absence of specular component) at the receiver, and there exist multiple reflective paths.[5] This is usually the case when the receiver (like cell phone) is situated in an urban environment or inside building complexes.

---

[3]Radio signals bounce back to the Earth even from the ionosphere.

[4]This type of noise is minimal when the transmitter is in space, or is airborne, and the receiver is in an open space like a desert, sea, or polar regions.

[5]The Rician distribution is preferred when there is a dominant LOS path along with few reflections.

18.6.2 WIND ENERGY MODELING

Rayleigh distribution is a better choice to model wind velocity in two dimensions, frequency of different wind speeds over a fixed time period.[6] Wind velocity is most easily analyzed by breaking it into its orthogonal 2-D vector components. The overall wind speed (vector magnitude) can be approximated by this distribution when each component is assumed to be uncorrelated Gaussians as $N(0, \sigma_k^2)$. The total number of hours during which wind will blow at speed v is given by

$$f(v) = B\pi/2(v/v_a^2)\exp(-(\pi/4)(v/v_a)^2), \qquad (18.21)$$

where B is the wind speed bin (e.g., 8760), v_a (or $\bar{v}$) is the average wind speed. The maximum of the above PDF occurs at $0.79788456v_a \approx 0.8v_a$, and the maximum energy wind is available at $1.6\,v_a$. The root-mean-cube measure ($\hat{v}_a = ((1/n)\sum_{k=1}^{n} x_k^3)^{1/3}$) is used in wind energy modeling to get the energy potential of a wind farm. A two-parameter Weibull(a, b) distribution is better when wind speed is needed for short time intervals like on a daily or weekly time scale, where the parameter a typically varies between 1.8 and 2.3. It reduces to the Rayleigh distribution for $a = 2$ (page 246). These are used for estimating energy recovery from wind turbines over a fixed time period. See [153] for an application of Rayleigh distribution to wind turbine modeling.

18.7 SUMMARY

This chapter introduced the Rayleigh distribution and its basic properties. This distribution can be obtained by size biasing half-normal distribution. Although it is a special case of χ^2 distribution, it finds applications in various engineering fields. It is increasingly being used in reliability engineering, wind-energy modeling, and communication sciences.

---

[6]Technically, we do not consider the direction of the wind, but only its speed which is positive (speed is zero when there is no wind). Different countries measure speed in different units (like nautical miles per hour). The unit used does not actually matter because scaled Rayleigh distributions belong to the same family. The International "small wind turbine standard" IEC 61400-2 recommends the scaled Rayleigh distribution, where the variate is scaled using the average wind speed $\bar{v}$.

CHAPTER 19

Maxwell Distribution

19.1 INTRODUCTION

This distribution is named after the Scottish physicist James Clerk Maxwell (1831–1879) whose work originated in kinetic theory of gases under equilibrium. The PDF is given by

$$f(x;a) = \sqrt{2/\pi}\; x^2 \exp(-x^2/(2a^2))/a^3 = (1/a)\sqrt{2/\pi}\;(x/a)^2 \exp(-\tfrac{1}{2}(x/a)^2),\;\; x > 0.$$

$$(19.1)$$

Here, "a" is a scale-parameter. Put $y = x/a$ (or put $a = 1$) to get the standard Maxwell distribution

$$f(x) = \sqrt{2/\pi}\;\; x^2 \exp(-x^2/2),\;\; x > 0. \tag{19.2}$$

It has a wide range of applications in diverse areas like thermodynamics, physical chemistry, nuclear and microwave engineering, etc., because it explains many fundamental properties of particles (atoms or molecules) in motion like speed, energies, and moments. It has been successfully applied in sedimentation of small particles in static low-density fluids. It is known as Maxwell–Boltzmann (MB) or Boltzmann–Maxwell (BM) distribution (in honor of German physicist Ludwig Boltzmann) in statistical mechanics, Gibbs distribution in particle physics, and Maxwellian distribution in thermal engineering. We denote it as Maxwell(a), or MB(a), and abbreviate it as MBD.

19.2 ALTERNATE FORMS

Put $2\sqrt{2}a/\pi = \mu$ to get a re-parametrized form

$$f(x;\mu) = (4/[\sqrt{\pi}C^3])x^2 \exp(-x^2/C^2),\;\; x > 0,\; C = (\pi\mu/2). \tag{19.3}$$

An alternate form in physics, known as Maxwell's velocity distribution (MVD), that represents the velocity of a gas molecule in thermal equilibrium is obtained under the substitution $a = \sqrt{kT/a}$ as

$$f(x;a,k,T) = 4\pi x^2(a/(2\pi kT))^{3/2} \exp(-ax^2/(2kT)), \tag{19.4}$$

where a = molecular weight, T = absolute temperature in Kelvin, and K = Boltzmann constant. Here, x is a scalar quantity. There is another popular representation in terms of a vector quantity as

$$dN/N = (m/(2\pi kT))^{1/2}x^2 \exp(-ax^2/(2kT))dv, \tag{19.5}$$

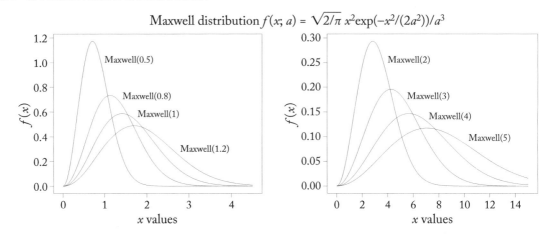

Figure 19.1: Maxwell distributions.

where dN/N is the fraction of molecules in the closed container moving at velocity between $(v, v + dv)$, and m is the mass of the molecule. This representation is useful to find out the number of molecules moving within the velocity differential $[v, v + dv]$. It is written in fluid mechanics in the form

$$f(x; n, T, v) = n/(2\pi T)^{3/2} \exp(-\frac{1}{2T}|x - v||x - v|'). \tag{19.6}$$

19.2.1   DERIVATION OF THE PDF

The following derivation assumes that a closed container of volume V has a finite number N of atoms (or molecules) moving at speed v in random directions, and colliding with each others continuously; moves frictionless, and do not dissipate energy via radiation or other means. Lots of gases behave as an ideal gas at very low pressure and high gas-specific temperatures. Consider a single molecule in 3D closed container with respective velocity components v_x, v_y, and v_z. It is assumed that the velocities are less than one-tenth of the speed of light (c). Since the speeds of the molecules are lower at small temperatures, the range of the distribution is short (see Figure 19.1). But as the temperature increase, the range also becomes large. As a few molecules (or atoms) happen to be at rest always, the curves are all anchored at zero (the origin). Ideally, this will not happen in practice at higher temperatures, in which case a left-truncated MBD may be more appropriate.

Each of these velocities change due to a large number of instantaneous collisions (with each other and with the surface of the enclosure). This is called *elastic collision* in particle physics, in which the rotational and vibrational kinetic energy is conserved. The resulting speed in a small epoch in time is given by $v = \sqrt{v_x^2 + v_y^2 + v_z^2}$. This shows that σ^2 is determined by the circular velocity. Due to symmetry, the mean velocity is $\sqrt{3\sigma^2}$. Kinetic energy in the x direction is $e_x = (m/2)v_x^2 = kT/2$

where T is the ambient temperature in Kelvin. As each velocity component contributes on the average $kT/2$, all three velocity components together contribute $3kT/2$ to the energy. Equating gives $3kT/2 = (m/2)(v_x^2 + v_y^2 + v_z^2) = (m/2)3\sigma^2$, where m is the mass of a molecule. Cancel (3/2) and simplify to get $kT = m\sigma^2$ or $\sigma = \sqrt{kT/m}$ as the standard deviation of velocity components.

To find the PDF of speed, consider an interval $(v, v + dv) = [(v_x, v_x + dv_x), (v_y, v_y + dv_y), (v_z, v_z + dv_z)]$. The PDF $f(v_x, v_y, v_z)$ is the probability per unit volume at an arbitrary point within the reach of molecules in space. Assuming that the perpendicular component velocities are statistically independent, the joint distribution of all the components depend only on functions of σ as $f(v_x, v_y, v_z) =$

$$f(v_x)f(v_y)f(v_z) = (1/C)\exp(-(v_x^2 + v_y^2 + v_z^2)/(2\sigma^2)) = (1/C)\exp(-v^2/(2\sigma^2)), \quad (19.7)$$

where $C = (2\pi)^{3/2}\sigma^3$. To get the PDF of speed, we have to multiply the above by the volume content of an infinitesimal spherical shell at radius v and thickness dv, which is approximately $4\pi v^2 dv$ (this is the surface area of a sphere of radius v). Drop the differential term dv to get

$$f(v) = (1/C)\exp(-v^2/(2\sigma^2)) \times 4\pi v^2 \quad = \sqrt{(2/\pi)}v^2/\sigma^3 \exp(-v^2/(2\sigma^2)). \quad (19.8)$$

Substitute for σ to get the classical Boltzmann distribution as a function of T. Alternately, write the joint PDF as

$$f(v_x, v_y, v_z) = (m/(2\pi kT))^{3/2} \exp(-m/(2kT)(v_x^2 + v_y^2 + v_z^2)). \quad (19.9)$$

Transform to spherical coordinates using $v_x = v\sin(\theta)\cos(\phi)$, $v_y = v\sin(\theta)\sin(\phi)$, and $v_z = v\cos(\theta)$, for which the Jacobian is $|J| = v^2\sin(\theta)$. The differential volume element $dx\,dy\,dz$ becomes $v^2\sin(\theta)dv d\theta d\phi$. The PDF of v can be found by integrating out θ and ϕ as

$$f(v) = \int_{\theta=0}^{\pi} \int_{\phi=0}^{2\pi} f(v_x, v_y, v_z)v^2\sin(\theta)v^2 d\phi d\theta. \quad (19.10)$$

Integrating gives

$$F(v) = 4\pi v^2(m/(2\pi kT))^{3/2}\exp(-mv^2/(2kT)). \quad (19.11)$$

This shows that raising the temperature causes the curve to skew to the right. In fact the root-mean-square speed and temperature are related as $T = mv^2/(3k)$. As $v^2 = 3kT/m$, the scale parameter measures the speed in units proportional to the square root of T/m (the ratio of temperature and particle mass). Physicists have experimentally proved that the molar mass distribution is very peaked (leptokurtic) when temperature is low or molecular mass is high, mesokurtic for intermediate values of them, and platykurtic for high temperature or low molecular mass. This has been extended to Doppler broadening of radiation frequency in which a static observer feels the resultant speed of molecules traveling in all directions (some of which is toward the observer, and some others away in opposite direction). Hence, the observed frequency is a combination of speeds and Doppler shifts.

The MBD has been found to be applicable in thermal vibrations and rotational states of particles (atoms or molecules) in solids, various spin-states in microwave engineering, electromagnetic radiation levels, and spectral diffusion (saturation transfer) in spectroscopy.

Although the upper limit of this distribution is ∞, it is truncated at the right in practice. The exact truncation point (for velocity distribution) depends on the ambiance. For example, models of gases in upper atmosphere of planets and moons use the respective escape velocity as upper-truncation point (this is 1.12×10^4 m/s for Earth). But, such a high truncation point does not matter much because the frequency curve goes down fast (X-axis that represents temperatures becomes an asymptote) much earlier, suggesting a small truncation point as shown in Figure 19.1. Moreover, ambiance typically contain a mixture of gases with various particle masses in different proportions for which the velocity distributions are different. One could also use mixtures of distributions with a "MB core" and more heavy-tailed non-MB distributions (like log-normal or Pareto) for large velocities.

A related distribution is that of the momentum of a particle (fraction of particles with momentum between p and $p + dp$) which is given by

$$f(p; m, k, T) = 4\pi/(2\pi mkT)^{3/2} p^2 \exp(-p^2/(2mkT))$$
$$= \sqrt{2/\pi}/(mkT)^{3/2} p^2 \exp(-p^2/(2mkT)). \qquad (19.12)$$

As the kinetic energy of a particle in motion is given by $E = mv^2/2$, we could derive the distribution of kinetic energy as

$$f(E; k, T) = 2\pi/(\pi kT)^{3/2} E^{1/2} \exp(-E/(kT))$$
$$= (2/\sqrt{\pi})/(kT)^{3/2} E^{1/2} \exp(-E/(kT)), \qquad (19.13)$$

which is independent of the molecular mass m. This is called Boltzmann distribution in thermodynamics. The spectral distribution of black-body radiation could also be derived from the above using the energy per unit volume $8\pi v^2 kT/c^3$ where c is the velocity of light.

19.3   RELATION TO OTHER DISTRIBUTIONS

It is the χ-distribution with three DoF. As such, it is related to the standard normal distribution as $X = (Z_1^2 + Z_2^2 + Z_3^2)^{1/2}$ where Z_k are IID $N(0, 1)$ random variables. We have seen the PDF of the normal (Gaussian) law $N(0, \sigma^2)$ as $f(x; 0, \sigma^2) = (1/(\sigma\sqrt{2\pi})) \exp(-x^2/(2\sigma^2))$. Note that the exponent involves x^2 scaled by $-2\sigma^2$. It was shown in Chapter 8 of Part I that $E(x^2) = \sigma^2$, which is the variance as $E(x) = 0$. Thus, the expected value of the exponent of Gaussian law is $-E(x^2)/(2\sigma^2) = -1/2$. Alternately, it follows from the integral $\int_0^\infty x^2 \exp(-ax^2) dx = \sqrt{\pi/a^3}$ or by differentiating $\int_0^\infty \exp(-bx^2) dx = .50\sqrt{\pi/b}$ wrt b on both sides.

As the χ_{n+2}^2 distribution is obtained by size-biasing a χ_n^2 distribution, it follows that the Maxwell distribution is the distribution of square-root of a size-biased χ_1^2 distribution. It can also be considered as a weighted folded (half-) Gaussian law $N(0, a^2)$ with weight x^2 (for Equation (19.1)) or $N(0, kT/m)$ for the speed version. The Arrhenius relationship used in reaction kinetics of physical chemistry uses the MB distribution. A change of origin transformation can be used to get a shifted version. Note that $f(x; a) = C \quad x^2 \exp(-Ax)$ is not a MBD but a quadratic weighted exponential distribution, which can easily be verified using gamma integral. Maxwell distribution is in fact a length-biased Rayleigh distribution. The distribution of $|X/Y|$ and $|XY|$ where X an Y are IID Maxwell and Rayleigh distributions can be found in Shakil, Golam, and Chang (2008) [199].

19.4   PROPERTIES OF MAXWELL DISTRIBUTION

The Maxwell and Rayleigh distributions are surprisingly similar-shaped for small parameter values. It is the limiting form of Fermi–Dirac distribution and Bose–Einstein distribution at high energies.

Problem 19.1   If the mass of a hydrogen atom on Earth is 3.34×10^{-27} kg, find the average speed of hydrogen atom on Earth surface where the temperature is 300K.

19.4.1   MOMENTS

Moments are easy to find using the integral $I_k = \int_0^\infty x^k \exp(-ax^2)dx$, which can be converted into gamma integral using the transformation $y = x^2$. In particular $I_0 = \sqrt{\pi}/(2\sqrt{a})$, $I_1 = 1/(2a)$. Mean $\mu = \int_0^\infty \sqrt{2/\pi} \quad x^2 \exp(-x^2/(2a^2))/a^3 dx$. Take constants outside the integral and put $y = x^2$ so that $dx = dy/(2\sqrt{y})$. Then $\mu = \sqrt{8/\pi}a = 2a\sqrt{2/\pi}$ and variance $\sigma^2 = (3 - 8/\pi)/a$. The mean velocity of a gaseous molecule at room temperature can then be estimated as $\bar{x} = [8kT/(a\pi)]^{1/2}$. Integration of (19.1) allows us to write the CDF in either of the following forms:

$$2\gamma(3/2, x^2/(2a^2))/\sqrt{\pi} = \text{erf}(x/(a\sqrt{2})) - (x/a)\sqrt{2/\pi}\exp(-x^2/(2a^2)). \qquad (19.14)$$

The fraction of the most energetic molecules whose energy state exceeds a specific threshold can be found using the SF.

Problem 19.2   Prove that the CDF of standard MBD can be expressed in terms of Gaussian CDF as $F(x) = 2\Phi(x) - \sqrt{2/\pi}\,x\exp(-x^2/2)$.

Problem 19.3   Prove that the points of inflection of standard MBD are at $\sqrt{(5 \mp \sqrt{17})/2}$.

The mode is $a\sqrt{2}$, and CV is $\sqrt{3\pi/8 - 1} \approx 0.422$. This is called the "most probable speed" in kinetic theory. The reliability function is $h(x) = 1 - 2/\sqrt{\pi}\gamma(3/2, ax^2)$. Parameter estimation for censored samples can be found in Arslan et al. (2017) [8]. Estimation of $\Pr[Y < X]$ for Maxwell distribution can be found in Chaudhary, Kumar, and Sanjeev (2017) [48].

Problem 19.4   Prove that the entropy of MBD is $\ln(a\sqrt{2\pi}) + \gamma - 0.5$ where $\gamma \approx 0.577215665$.

Table 19.1: Properties of Maxwell distribution $\sqrt{2/\pi}\ \ x^2 \exp(-x^2/(2a^2))/a^3$

Property	Expression	Comments
Range of X	$0 \leq x < \infty$	Continuous
Mean	$\mu = 2a\sqrt{2/\pi} = 1.595769\ a$	Median $\approx 1.53817\ a$
$E(X^2)$	$3a^2$	
Mode	$a\sqrt{2} = 1.414214\ a$	Mode $= 0.886227 *$ Mean
Variance	$\sigma^2 = (3 - 8/\pi)\ a^2$	$0.453521\ a^2$
Skewness	$2a\sqrt{2}(16 - 5\pi)/(3\pi - 8)^{3/2}$	$a * 0.4857$ approximately
MD	$4/\sqrt{\pi} \int_0^{2a\sqrt{2/\pi}} P(3/2, x^2/2a^2)dx$	$2\int_0^\mu P(3/2, x^2/2a^2)dx$
Moments	$\mu'_r = 2^{r/2+1}a^r\Gamma((r+3)/2)/\sqrt{\pi}$	$(r+1)!!a^r$, r even
CDF	$2\ P(3/2, x^2/2a^2))/\sqrt{\pi}$	
MGF	$2a\sqrt{2/\pi}\ t + \exp(a^2t^2/2)[2a^2t^2\Phi(at) + \sqrt{2/\pi}(1 - t)/a \exp(-t^2/2(a^2 - 1/a^2))]$	

Note that there are two parametrizations for Maxwell distribution.
mode < median < mean

Problem 19.5   Prove that the mean speed v_m, most probable speed v_o and root-mean-square speed v_r are related as $v_o < v_m < v_r$.

Problem 19.6   Find the temperature at which the average speed of H_2 atoms equals that of O_2 atoms at 200 K.

The MD is easily obtained using the power method as

$$\text{MD} = (4/\sqrt{\pi}) \int_0^{2a/\sqrt{\pi/2}} \gamma(3/2, x^2/(2a^2))dx. \tag{19.15}$$

This distribution is a special case of the chi-distribution which has PDF

$$f(x) = x^{n-1}e^{-x^2/(2\sigma^2)}/[2^{n/2-1}\sigma^n\Gamma(n/2)]. \tag{19.16}$$

See Table 19.1 for further properties.

The MLE of a is $\hat{a} = ((1/(3n))(\sum_{k=1}^n x_k^2))^{1/2}$, and the MoM estimate is $0.626657\ \bar{x}$.

Problem 19.7   Find the most probable speed of a particle of molecular weight .050 kg/mole at temperature 500 K.

Problem 19.8   Find the fraction of molecules with speed 1000 m/s that have molecular weight .050 kg/mole at T $= 1000$ K.

Problem 19.9   Prove that the energy distribution is given by $f(E) = 2\sqrt{E/\pi}\,(RT)^{-3/2}\exp(-E/(RT))$ where R is the gas constant (J/K) and $T =$ temperature in $K°$.

Problem 19.10   Find the energy distribution for a given temperature if the kinetic energy of a particle of molecular weight M is $E = .50Mc^2$ where c is the particle speed.

19.5   EXTENSIONS OF MAXWELL DISTRIBUTIONS

Truncated distributions are obtained by truncating in the left or right tail. The left-truncated Maxwell distribution has PDF

$$f(x; a, c) = \sqrt{2/\pi}\ \ x^2\exp(-x^2/(2a^2))/[a^3(1 - 2\,P(3/2, x^2/(2a^2))/\sqrt{\pi})], \ \ x > 0.$$
(19.17)

A size-biased Maxwell distribution can be obtained as in Chapter 1 of Part I as

$$f(x; a) = (1/(2a^4))x^3\exp(-x^2/(2a^2)), \quad x > 0,$$
(19.18)

with CDF

$$F(x; a) = 1 - (1 + x^2/(2a^2))\exp(-x^2/(2a^2)).$$
(19.19)

Alternatively, find the expected value of $(1 + cx)$ and proceed as in Chapter 1 of Part I. A transmuted Maxwell distribution has CDF $G(x; a, \lambda) = (1 + \lambda)F(x; a) - \lambda F^2(x; a)$. The corresponding PDF is $g(x; a, \lambda) = (1 + \lambda - 2\lambda F(x; a))f(x; a)$ where $\lambda \le 1$ is the shape parameter.

A power-Maxwell distribution is obtained by the transformation $Y = X^{1/b}$. This has PDF

$$f(x; a, b) = 4a^{3/2}b/\sqrt{\pi}\ \ x^{3b-1}\exp(-ax^{2b}), \quad x \ge 0,$$
(19.20)

where a is the scale and b is the shape parameter (Singh et al. (2018) [205], Segovia et al. (2020) [195]). This has mean $\mu'_k = (2/\sqrt{\pi})(1/a)^{k/(2b)}\Gamma((3b + k)/(2b))$. A transformation $Y = aX^{2b}$ can be used to find the CDF in terms of incomplete gamma function as $F(x; a, b) = P(ax^{2b}; 3/2)$. The kth raw moment is $E(x^k) = 2/(a^{k/(2b)})\Gamma((k + 3b)/(2b))$. The distribution of $Y = 1/X$ is called inverse Maxwell distribution. A review of inverse Maxwell distribution can be found in Tomer and Panwar (2020) [215].

Several other extensions of the MBD exist for specific purposes. For example, the Chapman-Enskog distribution is a perturbed MBD which is used in hybrid simulations of dilute gases.

Problem 19.11   If $f(r; a) = (r/a^2)\exp(-r^2/(2a^2))$ where $f(a) \sim \text{CUNI}(-\pi, \pi)$ find the unconditional distribution of r.

Problem 19.12   The number of atoms per unit volume of monatomic gas with energy in $[e, e + \delta e]$ is given by $f(e, T) = C/\sqrt{\pi}(kT)^{-3/2}\sqrt{e}\exp(-e/(kT))$ where K is Boltzman constant, T is the absolute temperature and e is the number of atoms. Find the unknown constant C and energy corresponding to the peak of the distribution.

19.6   APPLICATIONS

The MBD finds applications in nuclear engineering, physical chemistry, thermodynamics, and cryogenics, primarily in describing particle (atoms or molecules) speeds, or speed-ratio of idealized gases in thermodynamic equilibrium. Random motion of spherical particles in thermodynamics and molecular chemistry, and in vapor spectroscopy are approximated using the MBD. It is also used in sedimentation and diffusion studies involving motion of small particles in fluids, and suspended particle motion (like dust, smoke, aerosols, etc.) in media like non-uniform gases in closed containers or open atmosphere in colloidal chemistry to model the distribution of particle speeds exiting perpendicular to the surface of evaporating liquids in flat containers under thermal equilibrium, and reaction kinetics in chemical engineering. It has been successfully applied in efficient wet pump design and lubricant design in internal combustion (IC)-engines.

It is assumed that the density is homogeneous, temperature in an observational window is constant, the motion of a particle is chaotic, with speed never exceeding one-tenth of the speed of light, and that only binary encounters (single collisions) happen at any small time interval. These conditions are often met when the dilution (of the medium as in gaseous state) is small, and the temperature and pressure are conducive to chaotic motion in constrained boundary resulting in elastic collisions in which kinetic energy is not transferred. It is a popular choice in these fields due to its simplicity, ease of fitting, and flexibility, so that it can be used in IFR and DFR aging distributions. Nuclear and thermal engineers use it to approximate the distribution of neutron concentration with respect to energy (called neutron flux) in thermodynamic equilibrium within reactors, which is akin to the distribution of gas molecules described above. The distribution of radiation intensity, especially for *black body* can be approximated using the Rayleigh and Maxwell laws for specific ranges of the temperatures.[1] This approximation may not be suitable for low temperatures and high pressures as the distribution flattens out (dispersion increase rapidly).

19.6.1   KINETIC GAS THEORY

A gas in an enclosed container contains a large number of particles (atoms or molecules) in rapid random motion. These particles bombard with each other and the collision between them in turn changes the speeds. This of course depends on the chemical potential, temperature and pressure among other things. More and more particles attain greater kinetic energy as the temperature steadily increase (faster molecules collide more often than slower ones). The shape of the distribution flattens out at higher temperatures showing that particles move fast (have greater energy) at higher temperatures. Heavier molecules impart some of its kinetic energy to lighter ones if they are faster. It is assumed that neutrons are neither absorbed nor escape and the system of particles have reached a thermodynamic equilibrium[2] (only possible interaction is particle scattering). Maxwell–

---

[1]A black body is a hypothetical element that completely absorbs all electromagnetic radiation coming toward it, or equivalently whose electromagnetic frequency is below that of visible light.

[2]Thermodynamic equilibrium is reached when the energy of thermal motion and neutron energy are nearly equivalent so that the probabilities of gaining or losing of energy in a collision with nuclei are equal.

Boltzmann statistics denote energies of such particles in motion.[3] The distribution of particle speeds is derived by equating particle energy with kinetic energy. This results in the MBD in which the X-axis represents various speeds, and Y-axis represents the number (count) of particles moving at particular speeds. The most probable speed corresponds to the modal value of the distribution at $a\sqrt{2}$. As the MBD is uni-modal, the shape of the curve indicates that there are a large fraction of particles moving at maximum speed, and fewer number of them with smaller and larger speeds (indicated in the left and right side of the mode). Note that the shape of the distribution depends on the particle because the size of particles increase differently when temperature is raised. This means that those particles that increase in size when temperature goes up have greater chance of getting collided with each others than otherwise. This is the reason why MBD for hydrogen (H) and oxygen (O_2) are different.

It is well known that the mean molecular energy of gas in an enclosure is proportional to $(T/M)^{1/2}$, and the mean kinetic energy of thermal motion is $1.5\,kT$ where T is the temperature in Kelvin (K), k is the Boltzmann constant ($8.617333 \times 10^{-5}\,eV/K$ or equivalently $1.3806485 \times 10^{-23}\,J/K$), and M is the molecular mass. As per the Maxwell–Boltzmann law, the probability of a molecule occupying a kinetic energy E is proportional to $\exp(-E/kT)$ where T = temperature in K. For a particle in 3D, the speed and velocity vectors are related as $v = \sqrt{v_x^2 + v_y^2 + v_z^2}$.

19.6.2   CHEMISTRY

The rate at which chemical reactions take place in closed containers kept under equilibrium can be better understood using the MBD. Assume for simplicity that two chemicals (reactants) are combined and allowed to mix freely to produce a resulting chemical. This is symbolically denoted as $A + B \Rightarrow C$ where A, B are the reactants, and C is the resulting chemical. Usually the reactants A, B are at higher energy states than C. For this to happen, the molecules of A, B must collide with each other and overcome repulsive forces. Heating the container speeds up the process in most situations. This means that the temperature at which reaction takes place enhances the entire process because more molecules acquire greater amounts of kinetic energy at higher temperatures. Some chemical reactions are also enhanced using catalysts that act as conduits to help the molecules to collide. The 3D version of Maxwell distribution is used to model the process in closed containers.

19.7   SUMMARY

Maxwell distribution finds applications in thermodynamics, particle physics, physical chemistry, thermal engineering, and many other fields. It can be obtained by size-biasing Rayleigh distribution or as a weighted distribution of half-normal (with weight x^2). As it is a special case of Chi-distribution with three DoF discussed in Chapter 14, only important properties and applications not mentioned there are given here.

---

[3]The equilibrium distribution is Fermi–Dirac for electrons and Bose–Einstein for phonons, both of which converges to MB distribution at high energies.

CHAPTER 20

Gumbel Distribution

20.1 INTRODUCTION

The Gumbel distribution is named after Emil Julius Gumbel (1891–1966). This distribution is a member of the extreme value distributions (EVD),[1] which has a long history dating back to 1709 when Nicholas Bernoulli experimented with largest distance from the origin of n random numbers. There are three main types of EVD:

- type-I with CDF $F(x; a, b) = \exp(-\exp(-(x-a)/b))$;

- type-II with CDF $F(x; a, b, k) = \exp(-(b/(x-a))^k) = \exp(-((x-a)/b)^{-k})$, for $x \geq a, k \geq 2$, also called Frechet distribution; and

- type-III with CDF $F(x; a, b, k) = 1 - \exp(-((x-a)/b)^k)$, $x \geq 0$, which is the Weibull distribution.

EVDs are extensively used in failure analysis of structures and materials (like breaking strength), highways, communication infrastructures and utility systems, and breakdown voltage in electronic components (like inductors and capacitors). It is quite popular in applied sciences like climatology, hydrology, geology, seismology, and environmental sciences involving natural phenomena because the extreme values in the past data are of more interest in modeling optimal parameters. It has been used to model corrosion, wear and tear, etc., in civil engineering, reliability in structural engineering, extreme turbulent conditions in ocean engineering, gust velocities encountered in aeronautics (e.g., by aircrafts, drones, flying cars), and air pollution in environmental engineering. External materials of air-born crafts are so chosen as to withstand breaking strength caused by extreme gust conditions and vibrations. Engineering structures are often designed to withstand maximal load processes. The distribution of the maximum level of a dam or river can be modeled if reliable data on maximum values for several past years are available, using which structural engineers can predict the risk of exceedences. However, exceedences are measured on the nominal-scale (as counts like number of dam exceedences in a time period) or the ratio-scale in NOIR typology (Chattamvelli (2016) [41]).

Minimum extreme values are encountered in digital and wireless communications and reliability modeling, as in the case of failure of one-out-of-n serially connected identical components (this is called *weakest-link theory* in some fields). Similarly, the onset of drought and critical minimal water-levels in hydro-electric dams due to lack of enough rainfall depends on the minimum

---

[1]The distribution of largest or smallest data in a finite sample of size $n > 1$ is called extreme value.

rainfall received, which can be used to predict the corresponding intensity. Several industrial equipments, communication devices, and consumer products are also designed using minimum breaking strength of materials. Examples are the minimum thickness of insulators, or pipeline walls used in chemical and aerospace industries.[2] A continuous distribution with PDF given below is called a two-parameter Gumbel distribution:

$$f(x; a, b) = (1/b) \exp[-(x - a)/b - \exp(-(x - a)/b)], \quad -\infty < a, x < \infty. \quad b > 0. \quad (20.1)$$

Here "a" is the location and "b" the scale-parameter. Put $y = (x - a)/b$ (or put $a = 0, b = 1$) to get the standard Gumbel distribution (SGD)

$$f(x) = \exp(-x - \exp(-x)). \quad (20.2)$$

Consider a sample of size n from a population with PDF $f(x; \theta)$ and CDF $F(x; \theta)$ with $x < \infty$. The PDF of the maximum value (largest order statistic $x_{(n)}$) is given by

$$g(x; \theta) = n f(x; \theta)[F(x; \theta)]^{n-1}. \quad (20.3)$$

The Gumbel distribution results as the limiting form (as the sample size becomes large) of (20.3) where the upper limit of x is unrestricted. More precisely, it is the limiting form of $F(x)$ as $\lim_{n \to \infty}[F(a_n x + b_n)]^n$ for finite constants $a_n > 0$ and b_n. The Pareto, Cauchy, and log-gamma distributions converge to the aforementioned (Frechet) type II law, whereas normal, log-normal, gamma, etc., belong to type I. The PDF of the smallest element (smallest order statistic $x_{(1)}$) in a sample is obtained by replacing b by $-b$ inside the exponent in (20.1) as

$$f(x; a, b) = (1/b) \exp[(x - a)/b - \exp((x - a)/b)], \quad -\infty < a, x < \infty. \quad b > 0. \quad (20.4)$$

It is also called log-Weibull distribution or Fisher–Tippett distribution, and denoted as Gumbel(a, b), GU(a, b), or G(a, b). It is a particular case of the generalized Extreme Value distribution (GEVD) with CDF

$$F(x; a, b, c) = \exp[-(1 + (x - a)/(bc))^c], \quad -\infty < a, c, x < \infty. \quad b > 0. \quad (20.5)$$

20.1.1   ALTERNATE REPRESENTATIONS

The location parameter is denoted by μ and scale parameter by σ in some fields as

$$f(x; \mu, \sigma) = (1/\sigma) \exp(-(x - \mu)/\sigma - \exp(-(x - \mu)/\sigma)), \quad -\infty < x < \infty. \quad (20.6)$$

Put $c = 1/b$ to get

$$f(x; a, c) = c \exp(-c(x - a) - \exp[-c(x - a)]), \quad -\infty < a, x < \infty, \quad c > 0. \quad (20.7)$$

Take natural log of (20.1) to get

$$g(x; a, b) = \ln(f(x; a, b)) = (-\ln(b) - (x - a)/b - \exp(-(x - a)/b)), \quad -\infty < a, x < \infty. \quad (20.8)$$

---

[2]The space shuttle Challenger explosion of 1983 was caused by an extremely low temperature of 15° F.

20.2   RELATION TO OTHER DISTRIBUTIONS

It is most useful one when data comes from a distribution in which the right tail decay sub-exponentially. If X has a standard exponential distribution (SED), then $Y = -\ln(X)$ has an SGD.

Example 20.1   Exponential transformation of Gumbel (a, b)   If $X \sim$ Gumbel (a, b), then prove that $Y = \exp(-(x - a)/b) \sim$ SED.

Solution: From $y = \exp(-(x - a)/b)$ we have $x = a - b \log(y)$, so that the Jacobian is $|\partial x/\partial y| = |-b/y| = b/y$. Write the PDF as $(1/b) \exp[-(x - a)/b] \times \exp[-\exp(-(x - a)/b)]$, and put $y = \exp(-(x - a)/b)$ to get $f(y) = (1/b)y \exp(-y)$. Multiply this by the Jacobian to get $f(y) = \exp(-y)$, which is the SED.

Likewise, if $X \sim$ GU (a, b), then $Y = \exp(-X)$ is Weibull distributed. It is shown in Example 2.1 (page 19) of Chattamvelli and Shanmugam (2021) [47] that if X is a continuous random variable with CDF $F(x)$, then $U = F(x) \sim$ CUNI$(0, 1)$. As U and $1 - U$ are identically distributed, it follows that $V = S(x) = 1 - F(x) \sim$ CUNI$(0, 1)$. Thus, in general, $Y = -\log(S(x)) \sim$ SED for every continuous random variable, including Gumbel. The difference of two IID Gumbel-distributed random variables has a logistic distribution. More precisely, if $X \sim$ Gumbel (a, b) and $Y \sim$ Gumbel (c, b) are independent, then $Z = X - Y \sim$ Logistic$(a - c, b)$. If X and Y are IID Gumbel (a,b) then the distribution of their sum is approximately Logistic (2a, b).

The limiting distribution of Gumbel (a, b) is the Weibull distribution when the variate values are truncated (bounded from below) as in lifetime distributions. Another relationship exists among Weibull and extreme value distributions as follows. If $t_1, t_2, \ldots, t_n$ are random failure times from a Weibull distribution, then $\ln(t_1), \ln(t_2), \ldots, \ln(t_n)$ is a random sample from the extreme value distribution. Extreme values from Weibull and Cauchy distributions[3] are analyzed using statistics derived from exponential family of distributions using natural logarithmic transformation. Similarly, EVD of samples from normal and lognormal distributions have the same form, but with different parameters. The distribution of $Y = -X|Y > 0$ is called Gompertz distribution.

20.3   PROPERTIES OF GUMBEL DISTRIBUTION

The HF defined as $h(x) = -(\partial/\partial x) \log(S(x))$ has a simple form (it is called force of mortality in actuarial sciences). It is decreasing for $a < 1$, increasing for $a > 1$, and a constant for $a = 1$. For $n > 1$, the distribution of the maximum of a sample of size n is itself Gumbel distributed.

20.3.1   MOMENTS

The median of the Gumbel distribution with PDF $f(z) = \exp(-(z + \exp(-z)))$ is $-\ln(\ln(2)) \approx 0.3665129$, and that of (20.1) is $a - b \ln(\ln(2)) = a + 0.3665129b$, mean $\mu = a + b\gamma$ where $\gamma \approx$

---

[3]Note that the Cauchy distributions has range $(-\infty, \infty)$ so that logarithmic transformation is meaningful only for zero-truncated (half-Cauchy) distributions, but discussion is applicable for $x_{(n)}$ if the upper extreme is positive.

$$f(x, a, b) = (1/b) \exp(-(x - a)/b - \exp(-(x - a)/b))$$

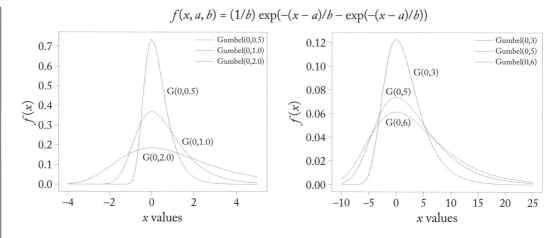

Figure 20.1: Gumbel distributions.

0.577215665 is Euler–Mascheroni constant, while the mode is a. The difference between mean and median is approximately $0.210687b$. It has variance $\pi^2 b^2/6 = 1.644934\, b^2$, which for the SGD becomes 1.6449341 (see Figure 20.1). The CV is $(\pi b/\sqrt{6})/(a + b\gamma) \approx 1.28254983/(a/b + 0.57722)$ or equivalently, $1/\text{CV} = 0.7796968\, a/b + 0.4500566$. It has points of inflections at $a \mp b \log((3 + \sqrt{5})/2) \approx a \mp 0.96242365\, b$.

Problem 20.2   If X and Y are IID Gumbel (a, b_i) for $i = 1, 2$, find the distribution of $U = X + Y$ and $V = X/Y$.

Problem 20.3   If $X \sim$ Gumbel (a, b), find the distribution of $U = floor(X)$ and $V = ceil(X)$.

20.3.2   ORDER STATISTIC

Let $x_1, x_2, \ldots, x_n$ denote a random sample of size n from a population with CDF $F(x)$. The kth order statistic is denoted by $x_{(k)}$, which is the kth item when the sample is arranged in ascending order. Let $X_{max} = max(x_1, x_2, \ldots, x_n)$ and $X_{min} = min(x_1, x_2, \ldots, x_n)$, The X_{min} and X_{max} are special cases corresponding to $k = 1$ and $k = n$. $F(max(X)) = \Pr[X_{max} \le x] = \Pr[X_1 \le x, X_2 \le x, \ldots X_n \le x]$. Due to independence, the RHS can be written as $\Pr[X_1 \le x] \Pr[X_2 \le x] \ldots \Pr[X_n \le x]$. As they are drawn from the same population, this becomes $F(x)^n$. Differentiate wrt x to get the PMF of maximum as $g(x) = n F(x)^{n-1} f(x)$ (see (20.3)). Similarly, the CDF of xmin is given by $F(min(X)) = 1 - (1 - F_1(x))(1 - F_2(x)) \ldots (1 - F_n(x)) = 1 - (1 - F(x))^n$. Differentiate wrt x to get the PDF of minimum as $g(x) = n(1 - F(x))^{n-1} f(x)$.

The CDF of the distribution of the largest element is

$$F(x; a, b) = \exp(-\exp[-(x - a)/b]), \quad -\infty < a, x < \infty. \ b > 0, \qquad (20.9)$$

and that of the smallest element is

$$F(x; a, b) = 1 - \exp[-(x - a)/b], \quad -\infty < a, x < \infty. \ b > 0, \tag{20.10}$$

from which the quantile function follows as $Q(p) = a - b \ln(-\ln(p))$. The above forms are used to get multi-parameter extensions of Gumbel distribution. The first and third quartiles are $Q_1 = a - 0.32663426 \, b$ and $Q_3 = a + 1.24589932 \, b$. The CDF is sometimes written as $F(x; a, b) = \exp(-\exp[-(cx + d)])$ where $c = 1/b, d = -a/b$.

Problem 20.4 Prove that the PDF evaluated at the mode is $1/(be) = 0.367879/b$.

Problem 20.5 Prove that the CDF evaluated at the mode is $1/e = 0.367879$, and at the median is $\exp(-\exp(\ln(\ln 2))) = 0.23629$.

If u_i is a uniform random number in the range $(0, 1)$, then random numbers from Gumbel(a, b) can be generated using $x_i = a - b \ln(\ln(1/u_i)) = a - b \ln(-\ln(u_i))$ (note that $-\ln(u_i)$ is always non-negative). Take natural log of (20.9) to get the relationship $-\ln(-\ln(F)) = (x - a)/b$, which can be regarded as a linearizing transformation. The resulting graph (in which Y-axis becomes the LHS) is called Gumbel plot. As this results in a straight line, these probability plots provide approximate estimates of a and b easily.

As a single logarithmic transformation results in exponential type, and double logarithmic transformation results in linear form, it is also called *doubly-exponential* distribution (whereas Laplace distribution (Chapter 13) is called *double-exponential*).[4]

Coefficient of skewness is $\mu_3/\sigma^3 = 1.13651$, and coefficient of kurtosis is $\mu_4/\sigma^4 = 5.40$. The MGF of GD is easy to obtain from the fact that $y = \exp(-(x - a)/b) \sim$ SED (see Example 20.1, page 265). Thus, $M_x(t) = \Gamma(1 - t)$, and that of Gumbel (a, b) is $M_x(t) = \Gamma(1 - bt) \exp(at)$. Estimation of $\Pr[Y < X]$ for Gumbel distribution can be found in [48]. See Table 20.1 for further properties.

Problem 20.6 The Gutenberg–Richter law of earthquake engineering has CDF $F(x; a, b) = (1 - \exp(-(x - a)/b))/1 - \exp(-(x_{max} - a)/b))$. Find the PDF and the mean.

20.4 FITTING

The MoM estimates are obtained by equating the first two moments as $\bar{x} = a + .57722 \, b$, and $s^2 = 1.64493406 \, b^2$ ($s = 1.28254983 \, b$). From these we get $\hat{b} = 0.7796968 \, s$, and $\hat{a} = \bar{x} - 0.4500566 \, s$. The MLE satisfies the equations

$$\hat{b} = \bar{x} - \sum_{j=1}^{n} x_j \exp(-x_j/\hat{b}) / \sum_{k=1}^{n} \exp(-x_k/\hat{b}) \tag{20.11}$$

$$\hat{a} = -\hat{b} \log(\sum_{j=1}^{n} \exp(-x_j/\hat{b})/n). \tag{20.12}$$

---

[4]Log-linear models are used in fatigue data analysis as an alternative to Birnbaum–Saunders distribution.

Table 20.1: Properties of Gumbel distribution $f(x;a,b) = (1/b)\exp[-(x-a)/b - \exp(-(x-a)/b)]$

Property	Expression	Comments
Range of X	$-\infty < x < \infty$	Continuous
Mean	$\mu = a + b\gamma$	$\gamma \approx 0.577215665$
Median	$a - b\ln(\ln(2))$	$a + 0.3665129b$
$E(X^2)$	$\gamma^2 + \pi^2/6$	1.978116995
Mode	a	
CV	$(\pi b/\sqrt{6})/(a + b\gamma) \approx 1.28255/(a/b + 0.57722)$	
Variance	$\sigma^2 = \pi^2 b^2/6$	1.644934 b^2
MD	$2\int_{-\infty}^{a+\gamma b} [\exp(-\exp[-(x-a)/b])]dx$	
CDF	$\exp(-\exp[-(x-a)/b])$	
Inv. CDF	$Q(p) = a - b\ln(-\ln(p))$	Quantile function
MGF	$M_x(t) = \Gamma(1 - bt)\exp(at)$	

Newton–Raphson method (NRM) can be used to solve the first equation, using which $\hat{b}$ is found, and substituted in the second equation to get $\hat{a}$. As the MoM estimates are inefficient, it can be used as an initial value for NRM. Alternately, use $b_0 = (\sqrt{6}/\pi)s = 0.7796968s$.

20.5   GENERALIZATION OF GUMBEL DISTRIBUTIONS

Truncated distributions are obtained by truncating in the left or right tail. The left-truncated Gumbel distribution has PDF

$$f(x;a,c) = (1/b)\exp[-(x-a)/b$$
$$- \exp(-(x-a)/b)]/[1 - \exp(-\exp[-(c-a)/b])], \quad x > c. \qquad (20.13)$$

A size-biased Gumbel distribution can be obtained as

$$f(x;a,b) = (1/b(a+b\gamma))x\exp[-(x-a)/b - \exp(-(x-a)/b)], \quad -\infty < x < \infty. \qquad (20.14)$$

Alternatively, find the expected value of $(1 + cx)$ and proceed as in Chapter 1 of Part I. A transmuted Gumbel distribution has CDF $G(x;a,\lambda) = (1 + \lambda)F(x;a) - \lambda F^2(x;a)$. It has PDF

$$g(x;a,\lambda) = \sqrt{2/\pi}(1 + \lambda - 2\lambda F(x;a))f(x;a). \qquad (20.15)$$

A power-Gumbel distribution is obtained by the transformation $Y = X^{1/c}$. This has PDF

$$g(y; a, b, c) = (c/b) \, f_x(y^c) \, \log(f_x(y^c)) \, y^{c-1}, \quad y \geq 0, \tag{20.16}$$

where a is the scale and b is the shape parameter. The distribution of $Y = 1/X$ is called inverse Gumbel distribution. See Nadarajah and Kotz (2004) [161] for beta-Gumbel distribution, Cordeiro et al. (2012) [52] for the Kumaraswamy–Gumbel distribution. Several other extensions of the Gumbel distribution exist for specific purposes. For example, the Chapman–Enskog distribution is a perturbed Gumbel distribution which is used in hybrid simulations of dilute gases. The Lomax–Gumbel distribution is obtained by convoluting the respective CDFs as $F_G(x) = F(G(x))/F(1)$

$$f(x; a, b, c, d) = K \left[1 - \{1 + (1/d \exp(-\exp((x-a)/b)))\}^{-c} \right], \quad x \geq 0 \tag{20.17}$$

$b, c, d > 0, -\infty < a < \infty$ (Gupta, Garg, and Gupta (2016) [85]).

20.6   APPLICATIONS

The study of stochastic behavior of events in the extreme tails of probability distributions is known as Extreme Value Theory (EVT). Gumbel distribution (with two or more parameters) is the most popular one in EVT. It has a wide range of applications in diverse areas like astronomy, disaster management, material science, statistical mechanics, etc. Average of an ensemble as well as extremes are of prime importance in statistical mechanics. It is a favorite choice in structural engineering and strength of materials. It is especially useful in engineering designs because extreme loads may have to be considered while designing structures like hydraulic infrastructures, cranes, bridges in flood-prone rivers, buildings in earthquake-prone regions, aircrafts and flying cars in gusty-wind situations, underwater robots in high-currents, and vehicles operating in extreme climatic conditions. It has been applied to predict the probability of extreme events like earthquakes, flood or other natural disasters. Automobile engineers use it to model mechanisms that lead to extreme events like breakdown of insulating fluids and engine oils, breakdown of various parts like transmission, clutch, wheels, etc.

It also finds applications in ocean and environmental engineering. Ocean engineers use it to predict the largest or strongest waves produced by natural phenomena like underwater earthquakes, volcano eruptions, etc. These are measured wrt a baseline (Mean Sea Level (MSL) for heights, and directions for waves in motion) for each such occurrence. Environmental engineers use it to predict extreme snowfalls, tornadoes, blizzard conditions, and other phenomena.

Although most of such calamities occur randomly, some periodicity is observed due to high-tide coinciding with such phenomena. Burke et al. (2010) [29] applied it to predict future changes in drought due to increased atmospheric greenhouse gases. The events of interest can be temporal, spatial, or a combination. Quite often a temporal event occurs along a spatial dimension as in earthquakes, floods, forest fires, etc. A bivariate model is more appropriate when two extreme events (e.g., average intensity and duration) are of interest. It relies on the principle that the magnitude of extreme events is related to their frequency of occurrence. In most practical applications, one is interested in modeling less likely (extreme) events. Mathematically, this is written as `magnitude = 1/(frequency of occurrence)`. Events are assumed to be independent and follow the same statistical distribution (usually exponential, Weibull, or normal law). Sometimes a cutoff threshold is used to reduce the size of the data. For example, the Peak-Over-Threshold (POT) method discards all data that are below a threshold. Such data are sometimes called POT data, and are guaranteed to include outliers. This acts as a high-pass filter to ease extreme value modeling on the high-end. A Goodness of fit (GoF) test can be performed to check the suitability of modeling when data comes from skewed distributions (Shanmugam(2013) [203]).

20.6.1   LAW OF RARE EXCEEDANCES

Several engineering structures and devices that transport fluids, liquids, and gases are designed to withstand a predetermined maximum load during its lifetime. Examples are canals, pipelines of various sorts (like water supply systems, oil and gas pipelines, feeds in automated factories like chemical and food processing industries, etc.), hydrolic and earthmoving equipments, etc. These are designed to withstand a maximum load (called critical load x_c) whose value is obtained from past data. The extra flow (load) beyond x_c is called an exceedance. It may occur due to natural causes (as in canals and dams due to rainfall), human errors, industrial factories due to over-heating, water supply systems due to malfunctions of pumps, electric power grids and electronic circuits due to voltage spikes, faulty sensors, etc. The probability that the system will not fail for n consecutive years is then given by $\Pr[X \leq x_c] = [F(x_c)]^n$ where X denotes the flow during a preset time interval (like peak flow in a day or during one operation cycle). Obviously, x_c is in the extreme right tail for modeling maximal flow. An exceedence X_α occurs when the flow during any unit time period exceeds the critical load x_c (so that $x_\alpha > x_c$). If $X \sim \text{Gumbel}(a, b)$, the probability that the system will sustain is given by

$$1 - Pr[X_\alpha] = (\exp(-\exp(-(x_c - a)/b)))^n, \qquad (20.18)$$

where a, b are estimated from the past data. Return period is the time interval between exceedances (in temporal events). The arithmetic mean of these time intervals is taken as the best estimate of return periods.

20.6.2   BIOINFORMATICS AND GENOMICS

Massive amounts of genomic data are automatically analyzed using statistical and machine learning algorithms. Sequence alignment algorithms are used to find commonality among two sequences to see how many features are similar among them. A score is obtained by the alignment of two real sequences a and b. Different score functions exist for this purpose. The maximal segment pair (MSP) is a technique used in matching proteins in bioinformatics and genes in genomics. These are converted into numeric scores, so that maximal matches can be identified. Extreme value distribution is one of the popular choices for MSP scores for random sequences, and in local alignments without gaps. The popular Lipman–Pearson model of genomics uses a Z-value computed from a random score distribution as $Z(a, b) = (s - \hat{\mu})/\hat{\sigma}$ where s is the score, $\hat{\mu}$ is the empirical mean, and $\hat{\sigma}$ is the standard deviation from a hypothetical distribution of scores $S(a, b)$, which is known to be approximately Gumbel (a, b).

20.7 SUMMARY

Gumbel distribution finds applications in many engineering fields. It is used to model "outlier" data values in either tails of a distribution, and life-time of devices, and parts. Electronic engineers use it to model endurance of components, and circuits. The minimum extreme value distribution is used when the left-tail of the life-time distribution increases at most exponentially, and maximum extreme value distribution when the right-tail decreases at least exponentially. Among the three types of EVD used in different contexts, the type-I EVD is the most popular in applied sciences. These are related through logarithmic transformations of the variates (Section 20.1, page 263). The CV in modeling problems involving maxima is quite large to justify the use of Gumbel distribution.

Bibliography

[1] Ahmed, S. E., Budsaba, K., Lisawadi, S., and Volodin, A. (2008). Parametric estimation for the Birnbaum–Saunders lifetime distribution based on new parametrization, distributions, *Thailand Statistician*, 6(2):213–240. https://www.researchgate.net/publication/237277945

[2] Alam, K. and Rizvi, H. M. (1967). On noncentral χ^2 and noncentral F distributions, *The American Statistician*, 67:21–22. https://www.jstor.org/stable/2682098, DOI: 10.1080/00031305.1967.10479830 204

[3] Almalki, S. J. and Nadarajah, S. (2014). A new discrete modified Weibull distribution, *IEEE Transactions on Reliability*, 63(1):166–180. DOI: 10.1109/tr.2014.2299691. 237

[4] Anderson, E. W. and Ellis, D. M. (1971). Error distributions in navigation, *Journal of the Institute of Navigation*, 24:429–442. DOI: 10.1017/s0373463300022281. 199

[5] Archer, C. O. (1967). *Some properties of Rayleigh distributed random variables and of their sums and products*, Naval Missile Center, Point Mugu, CA. DOI: 10.21236/ad0650090. 247

[6] Arellano-Valle, R. B., Gomez, H. W., and Quintana, F. A. (2005). Statistical inference for a general class of asymmetric distribution, *Journal of Statistical Planning and Inference*, 128:427–443. DOI: 10.1016/j.jspi.2003.11.014. 173

[7] Aroian, L. A. (1947). Note on the cumulants of Fisher's Z distribution, *Biometrika*, 34:359–360. DOI: 10.1093/biomet/34.3-4.359. 233

[8] Arslan, T., Acitas, S., and Senoglu, B. (2017). Parameter estimation for the two-parameter Maxwell distribution under complete and censored samples, *Revista Statistics*, 19(1):237–253. https://www.ine.pt/revstat/pdf/REVSTAT_v19-n2-04.pdf 257

[9] Awad, A. M. (1980). Remark on the characteristic function of the F-distribution, *Sankhya-A*, 42:128–129. https://www.ine.pt/revstat/pdf/REVSTAT_v19-n2-04.pdf 229

[10] Badarne, O. S., da Costa, D. B., Sofotasios, P. C., and Muhaidat, S. (2018). On the sum of Fisher–Snedecor F variates and its application to maximal-ratio combining, *IEEE Wireless Communications Letters*, 7(6):966–969, DOI: 10.1109/LWC.2018.2836453. 234

[11] Bailey, R. W. (1992). Distributional identities of beta and χ^2 variates: A geometric interpretation, *The American Statistician*, 46(2):117–120. DOI: https://doi.org/10.2307/2684178. 222

[12] Balakrishnan, N. and Kateri, M. (2008). On the maximum likelihood estimation of parameters of Weibull distribution based on complete and censored data, *Statistics and Probability Letters*, 78:2971–2975. DOI: 10.1016/j.spl.2008.05.019. 241

[13] Balakrishnan, N. and Kundu, D. (2019a). Birnbaum–Saunders distribution: A review of models, analysis, and applications (with discussions), *Applied Stochastic Models in Business and Industry*, 35(1):4–49. DOI: 10.1002/asmb.2348. 174

[14] Balakrishnan, N. and Kundu, D. (2019b). Birnbaum–Saunders distribution: A review of models, analysis and applications: Rejoinder, *Applied Stochastic Models in Business and Industry*, 35(1):126–132. DOI: 10.1002/asmb.2348. 172, 174

[15] Balakrishnan, N. and Nevzorov, V. B. (2003). *A Primer on Statistical Distributions*, John Wiley, NY. DOI: 10.1002/0471722227. 182, 222

[16] Balakrishnan, N. and Zhu, X. (2014). An improved method of estimation for the parameters of the Birnbaum–Saunders distribution, *Journal of Statistical Computation and Simulation*, 84:2285–2294. DOI: 10.1080/00949655.2013.789029. 172

[17] Barabesi, L. and Greco, L. (2002). A note on the exact computation of the Student-t, Snedecor-F and sample correlation coefficient distribution functions, *The Statistician*, 51:105–110, http://www.blackwellpublishing.com/content/bpl_images/Journal_Samples/ \RSSD0039--0526\51\1\294/302.PDF DOI: 10.1111/1467-9884.00302. 231

[18] Bardsley, W. E. (1980). Note on the use of the inverse Gaussian distribution for wind energy applications, *Journal of Applied Meteorology*, 19(9):1126–1135. DOI: 10.1175/1520-0450(1980)019%3C1126:notuot%3E2.0.co;2. 235

[19] Barreto, S. W., De-Morais A. L., and Cordeiro, G. M. (2011). The Weibull-geometric distribution, *Journal of Statistical Computation and Simulation*, 81(5):645–657. DOI: 10.1080/00949650903436554. 241

[20] Bhattacharyya, G. K. and Fries, A. (1982). Fatigue failure models—Birnbaum–Saunders vs. inverse Gaussian, *IEEE Transactions on Reliability*, 31:439–440. DOI: 10.1109/tr.1982.5221421. 167

[21] Birnbaum, Z. W. and Saunders, S. C. (1969a). A new family of life distributions, *Journal of Applied Probability*, 6:319–327. DOI: 10.1017/s0021900200032848. 163

[22] Birnbaum, Z. W. and Saunders, S. C. (1969b). Estimation for a family of life distributions with applications to fatigue, *Journal of Applied Probability*, 6:328–347. DOI: 10.1017/s002190020003285x. 172

[23] Bock, M. E. and Govindarajulu, Z. (1989). A note on the noncentral chi-square distribution, *Statistics and Probability Letters*, 7(2):127–129. DOI: 10.1016/0167-7152(88)90037-5. 205

[24] Boŝkoski, P. and Juriĉiĉ, D. (2014). Detection of bearing faults based on inverse Gaussian mixtures model, surveillance7.sciencesconf.org/conference/surveillance7/24_detection_of_bearing_faults_based_on_inverse_gaussian_mixtures_model.pdf 160

[25] Bottazzi, G., Cefis, E., and Dosi, G. (2002). Corporate growth and industrial structures: Some evidence from the Italian manufacturing industry, *Industrial and Corporate Change*, 11(4):705–723. DOI: 10.1093/icc/11.4.705. 195

[26] Bourguingon, M., Silva, R. B., Zea, L. M., and Cordeiro, G. M.(2013). The Kumaraswamy Pareto distribution, *Journal of Statistical Theory and Applications*, 12(2):129–144. DOI: 10.2991/jsta.2013.12.2.1. 187

[27] Bourguingon, M., Leao, J., Leiva V., and Santos-Neto, M. (2017). The transmuted Birnbaum–Saunders distribution, *REVSTAT—Statistical Journal*, 15:601–628. https://www.ine.pt/revstat/pdf/REVSTAT_v15-n4-8.pdf 173

[28] Brazauskas, V. and Kleefeld, A. (2011). Folded and log-folded-t distributions as models for insurance loss data, *Scandinavian Actuarial Journal*, pages 59–74. DOI: 10.1080/03461230903424199. 224

[29] Burke, E. J., Perry, R., and Brown, S. J. (2010). An extreme value analysis of UK drought and projections of change in the future, *Journal of Hydrology*, 388(1–2):131–143. DOI: 10.1016/j.jhydrol.2010.04.035. 269

[30] Cacoullos, T. (1964). A relation between t and F distributions, *Journal of American Statistical Society*, 60:528–531 and page 1249. DOI: 10.1080/01621459.1965.10480809. 217, 228

[31] Cameron, M. A. (1975). An extension of Ruben's result on the probability integral of non-central χ^2, *Communications in Statistics*, 4(8):783–786. DOI: 10.1080/03610927508827288. 205

[32] Casanova, P. S. and Agnana, T. C. (2000). About monotone regression quantiles, *Statistics and Probability Letters*, 48:101–104. DOI: 10.1016/S0167-7152(99)00200-X. 196

[33] Cassidy, D. T., Hamp, M. J., and Ouyed, R. (2010). Pricing European options with a log Student's t-distribution: A Gosset formula, *Physica A: Statistical Mechanics and its Applications*, 389(24):5736–5748. DOI: 10.1016/j.physa.2010.08.037. 217

[34] Castillo, N. O., Gomez, H. W., and Bolfarine, H. (2011). Epsilon Birnbaum–Saunder's distribution: Properties and inference, *Statistical Papers*, 52:871–883, DOI: 10.1007/s00362-009-0293-x. 173

[35] Chang, D. S. and Tang, L. C. (1994). Random number generator for the Birnbaum–Saunders distribution, *Computers and Industrial Engineering*, 27(1–4):345–348. DOI: 10.1016/0360-8352(94)90305-0.

[36] Chattamvelli, R. (1995a). A note on the noncentral beta distribution function, *The American Statistician*, 49:231–234. DOI: 10.1080/00031305.1995.10476151. 223, 232

[37] Chattamvelli, R. (1995b). Another derivation of two algorithms for the noncentral χ^2 and F distributions, *Journal of Statistical Computation and Simulation*, 49:207–214. DOI: 10.1080/00949659408811572. 204, 205

[38] Chattamvelli, R. (1996). On the doubly noncentral F distribution, *Computational Statistics and Data Analysis*, 20(5):481–489. DOI: 10.1016/0167-9473(94)00054-m. 232

[39] Chattamvelli, R. (2010). Power of the power-laws and an application to the PageRank metric, *PMU Journal of Computing Science and Engineering*, PMU, Thanjavur, TN 613403, India, 1(2):1–7.

[40] Chattamvelli, R. (2011). *Data Mining Algorithms*, Alpha Science, Oxford, UK. 242

[41] Chattamvelli, R. (2016). *Data Mining Methods*, 2nd ed., Alpha Science, Oxford, UK. 213, 263

[42] Chattamvelli, R. and Jones, M. C. (1995). Recurrence relations for noncentral density, distribution functions, and inverse moments, *Journal of Statistical Computation and Simulation*, 52(3):289–299. DOI: 10.1080/00949659508811679. 205, 206, 230

[43] Chattamvelli R. and Shanmugam R. (1995). Efficient computation of the noncentral χ^2 distribution, *Communications in Statistics—Simulation and Computation*, 24(3):675–689. DOI: 10.1080/03610919508813266. 205

[44] Chattamvelli, R. and Shanmugam, R. (1998). Computing the noncentral beta distribution function, Algorithm AS-310, *Applied Statistics*, Royal Statistical Society, 41:146–156. DOI: 10.1111/1467-9876.00055. 204

[45] Chattamvelli, R. and Shanmugam, R. (2019). *Generating Functions in Engineering and the Applied Sciences*, Morgan & Claypool. DOI: 10.2200/s00942ed1v01y201907eng037.

[46] Chattamvelli, R. and Shanmugam, R. (2020). *Discrete Distributions in Engineering and the Applied Sciences*, Morgan & Claypool. DOI: 10.2200/s01013ed1v01y202005mas034. 211

[47] Chattamvelli, R. and Shanmugam, R. (2021). *Continuous Distributions in Engineering and the Applied Sciences—Part 1*, Morgan & Claypool. DOI: 10.2200/s01076ed1v01y202101mas038. 158, 167, 169, 184, 186, 207, 220, 265

[48] Chaudhary, S., Kumar, J., and Sanjeev, K. T. (2017). Estimation of $\Pr[Y < X]$ for Maxwell distribution, *Journal of Statistics and Management Systems*, 20:467–481. https://www.researchgate.net/publication/319248752 257, 267

[49] Chhikara, R. S. and Folks, J. L. (1989). *The Inverse Gaussian Distribution: Theory, Methodology and Applications*, Marcel Dekker, NY. 155, 158

[50] Cohen, A. C., Whitten, B. J., and Ding, Y. (1984). Modified moment estimation for the three-parameter Weibull distribution, *Journal of Quality Technology*, 16:159–167. DOI: 10.1080/00224065.1984.11978908. 238

[51] Cordeiro, G. M. and Lemonte, A. J. (2010). The β- Birnbaum–Saunders distribution: An improved distribution for fatigue life modeling, *Computational Statistics and Data Analysis*, 55(3):1445–1461. DOI: 10.1016/j.csda.2010.10.007. 173

[52] Cordeiro, G. M., Nadarajah, S., and Ortega, E. M. M. (2012). The Kumaraswamy Gumbel distribution, *Statistical Methods and Applications*, 21:139–168. DOI: 10.1007/s10260-011-0183-y. 269

[53] D'Anna, G., Giorgio, M., and Riccio, A. (2016). Estimating fatigue life of structural components from accelerated data via a Birnbaum–Saunders model with shape and scale stress dependent parameters, *Procedia Engineering*, 167:10–17, Elsevier. www.sciencedirect.com DOI: 10.1016/j.proeng.2016.11.663. 176

[54] de Winter, J. C. F. (2013). Using the Student's t-test with extremely small sample sizes, *Practical Assessment, Research, and Evaluation*, 18(10). https://doi.org/10.7275/e4r6-dj05 https://scholarworks.umass.edu/pare/vol18/iss1/10 224

[55] den-Dekker, A. J. and Sijbers, J. (2014). Data distributions in magnetic resonance images: A review, *Physica Medica*, 30(7):725–741. DOI: 10.1016/j.ejmp.2014.05.002. 251

[56] De-Maesschalck, R., Jouan-Rimbaud, D., and Massart, D. L. (2000). The Mahalanobis distance, *Chemometrics and Intelligent Laboratory Systems*, 50(1):1–18. DOI: 10.1016/s0169-7439(99)00047-7. 213

[57] Desmond, A. F. (1986). On the relationship between two fatigue-life models, *IEEE Transactions on Reliability*, 35:167–169. DOI: 10.1109/tr.1986.4335393. 166, 167

[58] Dhaundiyal, A. and Singh, S. B. (2017). Approximations to the non-isothermal distributed activation energy model for biomass pyrolysis using the Rayleigh distribution, *Acta Tehnologica Agriculture*, 20:78–84. DOI: 10.1515/ata-2017-0016. 251

[59] Diaz-Garcia, J. A. and Dominguez-Molina, J. R. (2006). Some generalisations of Birnbaum–Saunders and sinh-normal distributions, *International Mathematical Forum*, 1(35):1709–1727. DOI: 10.12988/imf.2006.06146. 173

[60] Drapella, A. (1993). Complementary Weibull distribution: Unknown or just forgotten, *Quality and Reliability Engineering International*, 9:383–385. DOI: 10.1002/qre.4680090426. 241

[61] Edgeman, R. L. and Shanmugam, R. (1990). Linear regression analysis for inverse Gaussian processes: An illustration, *Decision Sciences Institute, Annual Meeting*, University of Georgia, Athens. 160

[62] Elgarhy, M., Shakil, M., and Golam-Kibria, B. M. (2017). Exponentiated Weibull-exponential distribution with applications, *Applications and Applied Mathematics*, 12(2):710–725. http://pvamu.edu/aam 241

[63] Eugene, N., Lee, C., and Famoye, F. (2002). Beta-normal distribution and its applications, *Communications in Statistics—Theory and Methods*, 31(4):497–512. DOI: 10.1081/sta-120003130. 173

[64] Fang, L., Zhu, X., and Balakrishnan, N. (2016). Stochastic comparisons of parallel and series systems with heterogeneous Birnbaum–Saunders components, *Statistics and Probability Letters*, 112:131–136. DOI: 10.1016/j.spl.2016.01.021. 176

[65] Finner, H., Dickhaus, T., and Roters, M. (2008). Asymptotic tail properties of Student's t-distribution, *Communications in Statistics—Theory and Methods*, 37(2):175–179. DOI: 10.1080/03610920701649019. 222

[66] Fisher, R. A. (1925). An expansion of Student's integral in powers of n^{-1}, *Metron*, 5:109–112. https://www.semanticscholar.org/ 222

[67] Fisher, R. A. (1931). The mathematical distributions used in the common tests of significance, *Econometrika*, pages 353–365. DOI: 10.2307/1905628. 204, 205

[68] Fisher, R.A. (1940). The precision of discriminant functions, *Annals of Eugenics*, 10:422–429, https://onlinelibrary.wiley.com/doi/pdf/10.1111/j.1469-1809.1940.tb02264.x. 233

[69] Fletcher, H. (1911). A verification of the theory of Brownian movements and a direct determination of the value of NE for gaseous ionization, *The Physical Review*, 33:81–110. DOI: 10.1103/physrevseriesi.33.81. 163

[70] Folks, J. L. and Chhikara, R. S.(1978) The inverse Gaussian distribution and its statistical application—a review, *Journal of the Royal Statistical Society-B*, 40(3):263–289. https://www.jstor.org/stable/2984691 DOI: 10.1111/j.2517-6161.1978.tb01039.x.

[71] Fox, W. R. and Lasker, G. W. (1983). The distribution of surname frequencies, *International Statistical Review*, 51(1):81–87. https://www.jstor.org/stable/1402733 DOI: 10.2307/1402733. 187

[72] Freudenthal, A. M. and Shinozuka, M. (1961). Structural safety under conditions of ultimate load failure and fatigue, *Technical Report TR-61-77*, Wright Patterson Air Force Base, Dayton, OH. 163

[73] From, G. S. and Li, L. (2006). Estimation of the parameters of the Birnbaum–Saunders distribution, *Communications in Statistics—Theory and Methods*, 35:2157–2169. DOI: 10.1080/03610920600853563. 171, 172

[74] Fujikoshi, Y. and Mukaihata, S. (1993). Approximations to the quantiles of Student's t and F distributions and their bounds, *Hiroshima Mathematical Journal*, 23:557–564. DOI: 10.32917/hmj/1206392782. 217

[75] Gabler, S. and Wolff, C. (1987). A quick and easy approximation to the distribution of a sum of weighted chi-square variables, *Statistiche Hefte*, 28:317–323. DOI: 10.1007/bf02932611. 203

[76] Gao, G., Wen, C., and Wang, H. (2017). Fast and robust image segmentation with active contours and Student's-t mixture model, *Pattern Recognition*, 63:71–86. DOI: 10.1016/j.patcog.2016.09.014. 225

[77] Garcia-Papani, F., Opazo, M. A., Leiva, V. and Aykroyd, R. G. (2017). Birnbaum–Saunders spatial modeling and diagnostics applied to agricultural engineering data, *Stochastic Environmental Research and Risk Assessment*, 31:105–124. DOI: 10.1007/s00477-015-1204-4. 176

[78] Garrett, R. G. (1989). The chi-square plot: A tool for multivariate outlier recognition, *Journal of Geochemical Exploration*, 32(1–3):319–341. DOI: 10.1016/0375-6742(89)90071-x. 213

[79] Gaunt, R. E. (2020). A simple proof of the characteristic function of Student's t-distribution, *Communications in Statistics—Theory and Methods*, https://arxiv.org/abs/1912.01245 DOI: 10.1080/03610926.2019.1702695. 218

[80] Gong, J., Lee, H., and Kang, J. (2020). Generalized MGF of inverse Gaussian distribution with applications to wireless communications, *IEEE Transactions on Vehicular Technology*, 69(2):2332–2336. DOI: 10.1109/tvt.2019.2962219. 158

[81] Good I. J. and Smith E. (1986). A power series for the tail-area probability of Student's T distribution, *Journal of Statistical Computation and Simulation*, 23(3):248–250. DOI: 10.1080/00949658608810876. 222

[82] Gordon, N. H. and Ramig, P. F. (1983). Cumulative distribution function of the sum of correlated chi-squared random variables, *Journal of Statistical Computation and Simulation*, 17:1–9. DOI: 10.1080/00949658308810633. 209

[83] Guiraud, P., Leiva, V., and Fierro, R. (2009). A non-central version of the Birnbaum–Saunders distribution for reliability analysis, *IEEE Transactions on Reliability*, 58:152–160. DOI: 10.1109/tr.2008.2011869.

[84] Guo, X., Wu, H., Li, G., and Li, Q. (2017). Inference for the common mean of several Birnbaum–Saunders populations, *Journal of Applied Statistics*, 44(5):941–954. DOI: 10.1080/02664763.2016.1189521. 172

[85] Gupta, J., Garg, M., and Gupta, M. (2016). The Lomax–Gumbel distribution, *Palestine Journal of Mathematics*, 5(1):35–42. https://pjm.ppu.edu/vol/2015/51 269

[86] Han, C. P. (1978). On the computation of noncentral chi-square distributions, *Journal of Statistical Computation and Simulation*, 6:207–210. DOI: 10.1080/00949657808810189. 205

[87] Han, C. P. (1979). On the computation of noncentral chi-square distributions with even degrees of freedom, *Journal of Statistical Computation and Simulation*, 9:25–29. DOI: 10.1080/00949657908810284. 205

[88] Harsaae, E. (1976). Some pitfalls in the use of minimum chi-square, *Statistische Hefte*, 17:81–104. DOI: 10.1007/bf02923059. 212

[89] Hartley, M. J. and Revankar, N. S. (1974). Estimation of the Pareto law from under-reported data, *Journal of Econometrics*, 5:1–11. DOI: 10.1016/0304-4076(77)90031-8. 198

[90] Harvey, A. and Lange, R. (2015). Volatility modeling with a generalized T distribution, *Cambridge Working Papers in Economics*, Faculty of Economics, Cambridge University, UK. 223

[91] Helmert F. R. (1876). Über die Wahrscheinlichkeit der Potenzsummen der Beobachtungs-fehler und uber einige damit in Zusammenhang stehende Fragen, *Zeitscrift der Mathematics und Physics*, 21:192–218. https://link.springer.com/journal/33/volumes-and-issues 201, 207, 215

[92] Hillery, M., O'Connell, R. F., Scully, M. O., and Wigner, E. P. (1984). Distribution functions in physics: Fundamentals, *Physics Letters*, 106(3):121–167. DOI: 10.1016/0370-1573(84)90160-1. 246

[93] Holla, M. S. and Bhattacharya, S. K. (1968). On a compound Gaussian distribution, *Annals of the Institute of Statistical Mathematics*, 20:331–336. DOI: 10.1007/bf02911647. 196

[94] Hribar, L. and Duka, D. (2010). Weibull distribution in modeling component faults, *IEEE Proceedings of ELMAR*, pages 183–186. https://ieeexplore.ieee.org/document/5606115 235

[95] Hsu, D. A. (1979). Long tailed distributions for position errors in navigation, *Journal of the Royal Statistical Society, Series C*, (Applied Statistics), 28:62–72. DOI: 10.2307/2346812. 199

[96] Iliescu, D. V. and Vodä, V. G. (1981). On the inverse Gaussian distribution, *Bulletin of Mathematical Sciences, Romania*, 25:281–293. https://www.jstor.org/journal/bullmathsocisci1 155

[97] Iriarte, Y. A., Gomez, H. W., Varela, H., and Bolfarine, H. (2015). Slashed Rayleigh distribution, *Revista Colombiana de Estadistica*, 38(1):31–44. https://revistas.unal.edu.co/ DOI: 10.15446/rce.v38n1.48800. 250

[98] James, I. R. (1979). Characterizations of a family of distributions by the independence of size and shape variables, *Annals of Statistics*, 7:869–881. DOI: 10.1214/aos/1176344736. 181

[99] Janssen, P. A. E. M. (2014). On a random time series analysis valid for arbitrary spectral shape, *Journal of Fluid Mechanics*, 79:236–256. DOI: 10.1017/jfm.2014.565. 251

[100] Janssen, P. A. E. M. (2015). Notes on the maximum wave height distribution, *European Center for Medium Range Weather Forecasts (ECMWF), Technical Report 755*, http://www.ecmwf.int/en/research/publications 251

[101] Jayalath, K. P. and Chhikara, R. S. (2020). Survival analysis for the inverse Gaussian distribution with the Gibbs sampler, *Journal of Applied Statistics*, DOI: 10.1080/02664763.2020.1828314. 160

[102] Jiang, L. and Wong, A. (2018). A chi-square approximation for the F distribution, *Open Journal of Statistics*, 8(1):146–158. https://www.scirp.org/journal/paperinformation.aspx?paperid=82593 DOI: 10.4236/ojs.2018.81010. 231

[103] Johnson N. L. (1959). On an extension of the connection between Poisson and χ^2 distributions, *Biometrika*, 46:352–363. DOI: 10.1093/biomet/46.3-4.352. 205

[104] Johnson, N. L., Kotz, S., and Balakrishnan, N. (2005). *Continuous Univariate Distributions*, vol. 1, 2, 2nd ed., John Wiley, NY. DOI: 10.1002/0471715816 172

[105] Jones, M. C. (1987). On the relationships between the Poisson-exponential model and the noncentral chi-squared distributions, *Scandinavian Actuarial Journal*, pages 104–109. DOI: 10.1080/03461238.1987.10413821 205

[106] Jones, M. C. (2002). On a class of distributions with simple exponential tails, *The Open University, Department of Statistics Technical Report 06/02*. https://www.open.ac.uk/TechnicalReports/jonesbka.pdf 190, 196

[107] Jones, M. C. (2008). On reciprocal symmetry, *Journal of Statistical Planning and Inference*, 138:3039–3043. DOI: 10.1016/j.jspi.2007.11.006. 167

[108] Jones, M. C. (2008b). The *t* family and their close and distant relations, *Journal of the Korean Statistical Society*, 37(4):293–302. DOI: 10.1016/j.jkss.2008.06.002. 217

[109] Jones, M. C. (2012). Relationships between distributions with certain symmetries, *Statistics and Probability Letters*, 82(9):1737–1744. DOI: 10.1016/j.spl.2012.05.014. 167

[110] Jones, M. C. (2019). On a characteristic property of distributions related to the Laplace, *South African Statistical Journal*, 53(1):31–34. 190

[111] Jones, M. C. and Faddy, M. J. (2003). A skew extension of the T distribution with applications, *Journal of Royal Statistical Society, Series B*, 65(1):159–174. DOI: 10.1111/1467-9868.00378. 224

[112] Kanji, G. K. (2006). *100 Statistical Tests*, Sage Publications. DOI: 10.4135/9781849208499. 224

[113] Kappenman, R. F. (1985). Estimation of the three-parameter Weibull, lognormal, and gamma distributions, *Computational Statistics and Data Analysis*, 3:11–23. DOI: 10.1016/0167-9473(85)90054-4. 238

[114] Keles, D. (2013). Uncertainties in energy markets and their consideration in energy storage evaluation, Dissertation, Karlsruhe Institute of Technology (KIT), Germany. www.ksp.kit.edu DOI: 10.5445/KSP/1000035365. 198

[115] Khan, A. H. and Jan, T. R. (2016). The inverse Weibull-geometric distribution, *International Journal of Modern Mathematical Sciences*, 14(2):134–146. www.ModernScientificPress.com/Journals/ijmms.aspx 241

[116] Khan, M. S. (2014). Modified inverse Rayleigh distribution, *International Journal of Computer Applications*, 87(13):28–43. DOI: 10.5120/15270-3868. 250

[117] Kirkby, J. L., Nguyen-Dang, and Nguyen-Duy (2019). Moments of student's t-distribution: A unified approach. DeepAI.org https://arxiv.org/pdf/1912.01607.pdf DOI: 10.2139/ssrn.3497188. 218

[118] Kocherlakota, K. and Kocherlakota, S. (1991). On the doubly noncentral *t* distribution, *Communications in Statistics—Simulation and Computation*, 20(1):23–31. DOI: 10.1080/03610919108812936. 223, 232

[119] Kotz, S., Johnson, N. L., and Boyd, D. W. (1967). Series representations of distributions of quadratic forms in normal variables-I: Central case, *Annals of Mathematical Statistics*, 38(3):823–837. DOI: 10.1214/aoms/1177698877. 203

[120] Kozubowski, T. J. and Inusah, S. (2006). A skew Laplace distribution on integers, *Annals of the Institute of Statistical Mathematics*, 58:555–571. DOI: 10.1007/s10463-005-0029-1. 190, 196, 237

[121] Kozubowski, T. J. and Inusah, S. (2006). *Multitude of Laplace Distributions*, Springer. DOI: 10.1007/s00362-008-0127-2. 195

[122] Kübler, H. (1979). On the fitting of the three-parameter distributions lognormal, gamma and Weibull, *Statistiche Hefte*, 20:68–125. DOI: 10.1007/BF02932451. 238

[123] Kundu, D., Kannan, N., and Balakrishnan, N. (2008). On the hazard function of Birnbaum–Saunders distribution and associated inference, *Computational Statistics and Data Analysis*, 52:2692–2702. DOI: 10.1016/j.csda.2007.09.021. 167

[124] Lai, C. D. (2014). *Generalized Weibull Distributions*, Springer. DOI: 10.1007/978-3-642-39106-4_2. 238, 241, 243

[125] Lai, J., Ji, D., and Yan, Z. (2020). Extended inverse Gaussian distribution: Properties and application, *Journal of Shanghai Jiaotong University*, (Science), 25:193–200. DOI: 10.1007/s12204-019-2144-9. 160

[126] Lawrence, J. (2013). Distribution of the median in samples from the Laplace distribution, *Open Journal of Statistics*, 3(6):41538. https://file.scirp.org/Html/6--1240248_41538.htm DOI: 10.4236/ojs.2013.36050. 193

[127] Lehmann, E. L. (2012). Student and small-sample theory, In *Selected Works of E. L. Lehmann*, pages 1001–1008, Springer, NY. DOI: 10.1007/978-1-4614-1412-4_83. 224

[128] Leiva, V. (2016). *The Birnbaum–Saunders distribution*, Elsevier/AP, Academic Press. DOI: 10.1016/C2014-0-04763-6. 171

[129] Leiva, V., Ponce, M. G., Marchant, C, and Bustos, O. (2012). Fatigue statistical distributions useful for modeling diameter and mortality of trees, *Revista Colombiana de Estadistica*, 35(3):349–370. https://revistas.unal.edu.co/ 173

[130] Leiva, V., Sanhueza, A., and Angulo, J. M. (2008). A length-biased version of the Birnbaum–Saunders distribution with application in water quality, *Stochastic Environmental Research and Risk Assessment*, Springer. DOI: 10.1007/s00477-008-0215-9.

[131] Leiva, V., Sanhueza, A., Sen, P. K. and Paula, G. A. (2008). Random number generators for the generalized Birnbaum–Saunders distribution, *Journal of Statistical Computation and Simulation*, 78(11):1105–1118. DOI: 10.1080/00949650701550242. 175

[132] Leão, J., Leiva, V., Saulo, H., and Tomazella, V. (2017). Birnbaum–Saunders fraility regression models: Diagnostics and application to medical data, *Biometrical Journal*, 59(2):291–314. DOI: 10.1002/bimj.201600008 176

[133] Leão, J., Saulo, H., Bourguignon, M., Cintra, R., Rêgo, L. and Cordeiro, G. (2013). On some properties of the beta inverse Rayleigh distribution, *Chilean Journal of Statistics*, 4(2):111–131. http://soche.cl/chjs/volumes/04/02/Leao_etal(2013).pdf 250

[134] Lemonte, A. J. (2013). A new extension of the Birnbaum–Saunders distribution, *Brazilian Journal of Probability and Statistics*, 27(2):133–149. DOI: 10.1214/11-bjps160. 173

[135] Li, B. and Martin, E. B. (2002). An approximation to the F distribution using the Chi-Square distribution, *Computational Statistics and Data Analysis*, 40:21–26. DOI: 10.1016/s0167-9473(01)00097-4. 231

[136] Li, R. and Nadarajah, S. (2020). A review of Student's *t* distribution and its generalizations, *Empirical Economics*, 58:1461–1490, Springer. DOI: 10.1007/s00181-018-1570-0. 217, 224

[137] Lillo, C., Leiva, V., Nicolis, O., and Aykroyd, R. G. (2018). L-moments of the Birnbaum–Saunders distribution and its extreme value version: Estimation, goodness of fit and application to earthquake data, *Journal of Applied Statistics*, 45(2):187–209. DOI: 10.1080/02664763.2016.1269729. 175

[138] Lin, J. T. (1988). Approximating the cumulative chi-square distribution and its inverse, *Statistician*, 37:3–5. DOI: 10.2307/2348373. 207

[139] Liu, H., Tang, Y., and Zhang, H. H. (2009). A new chi-square approximation to the distribution of non-negative definite quadratic forms in non-central normal variables, *Computational Statistics and Data Analysis*, 53:1–8. DOI: 10.1016/j.csda.2008.11.025. 203

[140] Luo, L., Fang, C. et al. (2018). Study on probability distribution of photovoltaic power fluctuations at multi-time scales, *CIRED Workshop-Ljubljana*, paper 0266. DOI: 10.34890/441. 195

[141] Lúroth, J. (1876). Vergleichung von zwei Werten des wahrscheinlichen Fehlers, *Astronomice Nachrichten*, 87(14):209–220. DOI: 10.1002/asna.18760871402. 215

[142] Maehara, R., Bolfarine, H., Vilca, F., and Balakrishnan, N. (2021). A robust Birnbaum–Saunders regression model based on asymmetric heavy-tailed distributions, *Metrika*, DOI: 10.1007/s00184-021-00815-4. 167, 176

[143] Marchant, C., Leiva, V., Cavieres, M. F., and Sanhueza, A. (2013). Air contaminant statistical distributions with application to PM10 in Santiago, Chile, *Reviews of Environmental Contamination and Toxicology*, 223:1–31. DOI: 10.1007/978-1-4614-5577-6_1. 173

[144] Mardia, K. V. (1974). Applications of some measures of multivariate skewness and kurtosis in testing normality and robustness studies, *Sankhya-B*, 36:115–128. 203

[145] Martinez-Florez, G., Barranco-Chamorro, I., Bolfarine, H., and Gomez, H. W. (2019). Flexible Birnbaum–Saunders distribution, *Symmetry*, 11:1305. DOI: 10.3390/sym11101305. 167

[146] Mason, I. G., McLachlan, R. I., and Gerard, D. T. (2006). A double exponential model for biochemical oxygen demand, *Bioresource Technology*, 97:273–282. sciencedirect.com DOI: 10.1016/j.biortech.2005.02.042. 198

[147] Mathai, A. M. and Provost, S. B. (1992). *Quadratic Forms in Random Variables—Theory and Applications*, Marcel Dekker, NY. DOI: 10.2307/2290674. 207

[148] McDonald, J. B. and Newey, W. K. (1988). Partially adaptive estimation of regression models via the generalized *t* distribution, *Econometric Theory*, 4(3):428–457. https://www.jstor.org/stable/3552334. 223

[149] McEwen R. P. and Parreson, B. R. (1991). Moment expressions and summary statistics for the complete and truncated Weibull distributions, *Communications in Statistics—Theory and Methods*, 20:1361–1372. DOI: 10.1080/03610929108830570. 241

[150] McFadyen, A. and Martin, T. (2018). Understanding vertical collision risk and navigation performance for unmanned aircraft, *Proc. of 37th AIAA/IEEE Digital Avionics Systems Conference*, pages 1–10. https://eprints.qut.edu.au/124768/ DOI: 10.1109/dasc.2018.8569707. 199

[151] Meitz, M., Preve, D. P. A., and Saikkonen, P. (2018). A mixture autoregressive model based on student's t-distribution. https://ssrn.com/abstract=3177419 DOI: 10.1080/03610926.2021.1916531. 225

[152] Menon, M. V. (1963). Estimation of the shape and scale parameters of the Weibull distributions, *Technometrics*, 5:175–182. DOI: 10.1080/00401706.1963.10490073. 238

[153] Mihet-Popa, L. and Groza, V. (2011). Annual wind and energy loss distribution for two variable speed wind turbine concepts of 3 MW, *Instrumentation and Measurement Technology Conference (I2MTC)*, IEEE, pages 1–5. DOI: 10.1109/imtc.2011.5944340. 252

[154] Mittal, M. M. and Dahiya, R. C. (1989). Estimating the parameters of a truncated Weibull distribution, *Communications in Statistics—Theory and Methods*, 18:2027–2042. DOI: 10.1080/03610928908830020. 241

[155] Mohammadi, K., Alavi, O., and McGowan, G. (2017). Use of Birnbaum–Saunders distribution for estimating wind speed and wind power probability distributions: A review, *Energy Conversion and Management*, 143(1):109–122. DOI: 10.1016/j.enconman.2017.03.083. 175

[156] Mudholker, G. S. and Natarajan, R. (2002). The inverse Gaussian models: Analogues of symmetry, skewness and kurtosis, *Annals of Institute of Statistical Mathematics*, 54(1):138–154. DOI: 10.1023/A:1016173923461. 158

[157] Mudholkar, G. S. and Srivastava, D. K. (1993). Exponentiated Weibull family for analyzing bathtub failure-rate data, *IEEE Transactions on Reliability*, 42(2):299–302. DOI: 10.1109/24.229504. 241

[158] Mukherjee, S. P. and Sasmal, B. C. (1984). Estimation of Weibull parameters using fractional moments, *Calcutta Statistical Association Bulletin*, 33:179–186. DOI: 10.1177/0008068319840308. 238

[159] Murthy, D. N. P., Xie, M., and Jiang, R. (2003). *Weibull Models*, John Wiley, NY. DOI: 10.1002/047147326x. 235, 241

[160] Nadarajah, S. (2005). Exponentiated Pareto distributions, *Statistics*, 39(3):255–260. DOI: 10.1080/02331880500065488. 187

[161] Nadarajah, S. and Kotz, S. (2004). The beta-Gumbel distribution, *Mathematical Problems in Engineering*, 4(1):323–332. DOI: 10.1155/s1024123x04403068. 269

[162] Nadarajah, S. and Kotz, S. (2006). The exponentiated type distributions, *Acta Applied Mathematics*, 92:97–111. DOI: 10.1007/s10440-006-9055-0. 241

[163] Nar, F., Okman, O. E., Özgür, A., and Cetin, M. (2018). Fast target detection in radar images using Rayleigh mixtures and summed area tables, *Digital Signal Processing*, 77:86–101. DOI: 10.1016/j.dsp.2017.09.015. 251

[164] Narula, S. C. and Franz, S. L. (1977). Approximations to the chi-square distribution, *Journal of Statistical Computation and Simulation*, 5:267–277. DOI: 10.1080/00949657708810157. 207

[165] Nassar, M., Afify, A. Z., Dey, S., and Kumar, D. (2018). A new extension of Weibull distribution: Properties and different methods of estimation, *Journal of Computational and Applied Mathematics*, 336:439–457. DOI: 10.1016/j.cam.2017.12.001. 241

[166] Ng, H. K. T., Kundu, D., and Balakrishnan, N. (2003). Modified moment estimation for the two-parameter Birnbaum–Saunders distribution, *Computational Statistics and Data Analysis*, 43(3):283–298. DOI: 10.1016/s0167-9473(02)00254-2. 172

[167] Oliveira, K. L. P., Castro, B. S., Saulo, H., and Vila, R. (2020). On a length-biased Birnbaum–Saunders regression model applied to meteorological data, *ArXiv:2012.10760v*. 173, 176

[168] Owen, W. J. (2006). A new three-parameter extension to the Birnbaum–Saunders distribution, *IEEE Transactions on Reliability*, 55:475–479. DOI: 10.1109/tr.2006.879646. 173

[169] Paula, G. A., Leiva, V., Barros, M., and Liu, S. (2011). Robust statistical estimations using the Birnbaum–Saunders-t distribution applied to insurance, *Applied Stochastic Models in Business and Industry*, 28(1):16–34. DOI: 10.1002/asmb.887. 177

[170] Pearson, K. (1895). Contributions to the mathematical theory of evolution II—skew variation in homogeneous material, *Philosophical Transactions of the Royal Society-A*, 186:343–414 (374). 215

[171] Pescim, R. R. et al. (2014). The Kummer beta Birnbaum–Saunders: An alternative fatigue life distribution, *Hacettepe Journal of Mathematics and Statistics*, 43(3):473–510. http://www.hjms.hacettepe.edu.tr/ 173, 175

[172] Pestana, D. (1977). Note on a paper of Ifram, *Sankhya-A*, 39:396–97. 229

[173] Phillips, P. C. B. (1982). The true characteristic function of the F distribution, *Biometrika*, 69:261–264. DOI: 10.1093/biomet/69.1.261. 229

[174] Platen, E. and Sidorowicz, R. (2007). Empirical evidence on Student-t log-returns of diversified world stock indices, *Research Paper 194, Quantitative Finance Research Centre*, University of Technology, Sydney, Australia. http://www.business.uts.edu.au/qfrc/research/research_papers/rp194.pdf DOI: 10.1080/15598608.2008.10411873. 217

[175] Pook, L. (2007). *Metal Fatigue: What it is, Why it Matters*, Springer. DOI: 10.1007/978-1-4020-5597-3. 240

[176] Probst, A. C., Braun, M., Backes, J., and Tenbohlen, S. (2011). Probabilistic analysis of voltage bands stressed by electric mobility, *2nd IEEE PES International Conference on Innovative Smart Grid Technologies*. DOI: 10.1109/isgteurope.2011.6162773. 187

[177] Punzo, A. (2019). A new look at the inverse Gaussian distribution with applications to insurance and economic data, *Journal of Applied Statistics*, 46(7):1260–1287. https://arxiv.org/pdf/1707.04400.pdf DOI: 10.1080/02664763.2018.1542668. 160

[178] Qu, H. and Xie, F. (2011). Diagnostic analysis for log-Birnbaum–Saunders regression models with censored data, *Statistica Neerlandica*, 65(1):1–21. DOI: 10.1111/j.1467-9574.2010.00467.x. 176

[179] Quandt, R. E. (1966). Old and new methods of estimation and the Pareto distribution, *Metrika*, 10:55–82. DOI: 10.1007/bf02613419. 184

[180] Rao, C. R. (1973). *Linear Statistical Inference and its Applications*, 2nd ed., John Wiley, NY. 207

[181] Reyes, J. et al. (2018). Generalized modified slash Birnbaum–Saunders distribution, *Symmetry*, 10(12):724. DOI: 10.3390/sym10120724.

[182] Rieck, J. R. (1999). A moment generating function with applications to Birnbaum–Saunders distribution, *Communications in Statistics—Theory and Methods*, 28:2213–2222. DOI: 10.1080/03610929908832416. 169

[183] Rieck, J. R. (2003). A comparison of two random number generators for the Birnbaum–Saunders distribution, *Communications in Statistics—Theory and Methods*, 32:929–934. DOI: 10.1081/sta-120019953. 175

[184] Rieck, J. R. and Nedelman, J. R. (1991). A log-linear model for the Birnbaum–Saunders distribution, *Technometrics*, 33:51–60. DOI: 10.1080/00401706.1991.10484769. 167

[185] Robert, C. (1990). On some accurate bounds for the quantiles of a noncentral chi-squared distribution, *Statistics and Probability Letters*, 10:101–106. DOI: 10.1016/0167-7152(90)90003-p. 207

[186] Rosenberg, L. and Bocquet, S. (2015). Application of the Pareto plus noise distribution to medium grazing angle sea-clutter, *IEEE Journal of Selected Topics in Applied Earth Observations and Remote Sensing*, 8(1):255–261. DOI: 10.1109/jstars.2014.2347957. 187

[187] Rosen, P. and Rammler, E. (1933). The laws governing the fineness of powdered coal, *Journal of the Institute of Fuel*, 7:29–36. 235

[188] Ruben, H. (1974). A new result on the probability integral of the noncentral chi-squared with even degrees of freedom, *Communications in Statistics—Theory and Methods*, 3(5):473–476. DOI: 10.1080/03610927408827148 202, 205

[189] Salo, J., El-Sallabi, M., and Vainikainen, P. (2006). The distribution of the product of independent Rayleigh random variables, *IEEE Transactions on Antenna and Propagation*, 54(2):639–643. DOI: 10.1109/tap.2005.863087. 247

[190] Santos-Neto, M. et al. (2014). A reparameterized Birnbaum–Saunders distribution and its moments, estimation and applications, *REVSTAT—Statistical Journal*, 12(3):247–272. https://www.ine.pt/revstat/pdf/rs140303.pdf 174

[191] Santos-Neto, M. et al. (2016). A reparameterized Birnbaum–Saunders regression model with varying precision, *Electronics Journal of Statistics*, 10:2825–2855. https://projecteuclid.org/journals/electronic-journal-of-statistics/volume-10/issue-2/ 176

[192] Sarti, A., Corsi, C., Mazzini, E., and Lamberti, C. (2005). Maximum likelihood segmentation of ultrasound images with Rayleigh distribution, *IEEE Transactions on Ultrasound*, 52:947–960. DOI: 10.1109/tuffc.2005.1504017. 251

[193] Sato, S. and Inoue, J. (1994). Inverse Gaussian distribution and its application, *Electronics and Communications in Japan (Part III)*, 77(1):32–42. DOI: 10.1002/ecjc.4430770104.

[194] Saulo, H., Leao, J., and Bourguignon, M. (2012). The Kumaraswamy Birnbaum–Saunders distribution, *Journal of Statistical Theory and Practice*, 6(4):745–759. DOI: 10.1080/15598608.2012.719814. 173

[195] Segovia, F. A., Gomez, Y. M., Venegas, O., and Gomez, H. W. (2020). A power Maxwell distribution with heavy tails and applications, *Sigma Mathematics*, 8:1–20. mdpi.com/journal/mathematics DOI: 10.3390/math8071116. 259

[196] Sen, P. K. (1989). The mean-median-mode inequality and noncentral chi-square distributions, *Sankhya A*, 51:106–114.

[197] Sengupta, S. (1991). Some simple approximations for the doubly noncentral z distribution, *Australian Journal of Statistics*, 33:177–181. DOI: 10.1111/j.1467-842x.1991.tb00425.x. 232

[198] Seshadri, V. (1999). *The Inverse Gaussian Distribution: Statistical Theory and Applications*, Springer, NY. DOI: 10.2307/2670031. 155, 161

[199] Shakil, M., Golam, B., and Chang, K. (2008). Distribution of the product and ratio of Maxwell and Rayleigh random variables, *Statistical Papers*, 49:729–747. DOI: 10.1007/s00362-007-0052-9. 257

[200] Shanmugam, R. (1987). Estimating the fraction of population in an income bracket using Pareto distribution, *Brazilian Journal of Probability and Statistics*, 1:139–156. 184, 187, 234

[201] Shanmugam, R. (1999). Testing the homogeneity of a random sample from Pareto distribution, *Biometrie Praximetrie*, 29(2):75–91. 184, 187

[202] Shanmugam, R. (2009). A tutorial of diagnostic methodology with 2 x 2 dementia data, *International Journal of Data Analysis Techniques and Strategies*, 2:385–406. DOI: 10.1504/IJDATS.2009.027516 214

[203] Shanmugam, R. (2013). Alternate to traditional goodness of fit test with illustration using service duration to patients in hospitals, *International Journal of Statistics and Economics*, 11(2):31–43. http://www.ceser.in/ceserp/index.php/bse/article/view/2144/2771 270

[204] Shanmugam, R. (2020). Does how long observing correlate with upper record values?, *Communications in Statistics—Theory and Methods*. DOI: 10.1080/03610926.2020.1815783. 245

[205] Singh, A., Bakouch, H., Kumar, S., and Singh, U. (2018). Power Maxwell distribution: Statistical properties, estimation and applications, *ArXiv:1808.01200v1*. 259

[206] Singh, K. P. and Lee, C. M. S. (1988). On the t cumulative probabilities, *Communications in Statistics—Simulation and Computation*, 17(1):129–135. DOI: 10.1080/03610918808812652. 222

[207] Smith, K. (1916). On the "best" values of the constants in frequency distributions, *Biometrika*, 11:262–276. DOI: 10.1093/biomet/11.3.262. 212

[208] Sob, U. M., Bester, H. L., Smirnov, O. M., Kenyon, J. S., and Grobler, T. L. (2020). Radio interferometric calibration using a complex student's t-distribution and Wirtinger derivatives, *Monthly Notices of the Royal Astronomical Society*, 491(1):1026–1042, DOI: 10.1093/mnras/stz3037. 225

[209] Sousa, M. F. (2016). Two essays on Birnbaum–Saunders regression models for censored data, Masters thesis, Universidade Federal de Goias (UFG), Brazil. https://repositorio.bc.ufg.br/tede/handle/tede/7235 176

[210] Student (1908). The probable error of the mean, *Biometrika*, 6:1–25. https://www.jstor.org/stable/2331554 215

[211] Takagi, K., Kumagai, S., Matsunaga, I., and Kusaka, Y. (1997). Application of inverse Gaussian distribution to occupational exposure data, *Annals of Occupational Hygiene*, 41(5):505–514. DOI: 10.1016/s0003-4878(97)00015-x. 155

[212] Talha, B. and Pätzold, M. (2007). Statistical modeling and analysis of mobile-to-mobile fading channels in cooperative networks under line-of-sight conditions, *Wireless Personal Communications*, 54:3–19, Springer. DOI: 10.1007/s11277-009-9721-4. 198

[213] Teimouri, M., Hosseini, S. M., and Nadarajah, S. (2013). Ratios of Birnbaum–Saunders random variables, *Quality Technology and Quantitative Management*, 10(4):457–481. DOI: 10.1080/16843703.2013.11673425. 173

[214] Tiku, M. L. (1971). Students t distributions under non-normal situations, *Australian Journal of Statistics*, 13(3):142–148. DOI: 10.1111/j.1467-842X.1971.tb01253.x. 222

[215] Tomer, S. K. and Panwar, M. S. (2020). A review on inverse Maxwell distribution with its statistical properties and applications, *Journal of Statistical Theory and Practice*, 14(33). DOI: 10.1007/s42519-020-00100-z. 259

[216] Trigui, I., Laourine, A., Affes, S., and Stéphenne, A. (2012). The inverse Gaussian distribution in wireless channels: Second-order statistics and channel capacity, *IEEE Transactions on Communication*, 60(11):3167–3173. DOI: 10.1109/tcomm.2012.081512.100253.

[217] Tronarp, F., Karvonen, T., and Särkkä, S. (2019). Student's t-filters for noise scale estimation, *IEEE Signal Processing Letters*, 26(2):352–356. DOI: 10.1109/LSP.2018.2889440. 224

[218] Tweedie M. C. K. (1947). Functions of a statistical variate with given means, with special reference to Laplacian distributions, *Proc. of the Cambridge Philosophical Society*, 43:41–49. DOI: 10.1017/s0305004100023185. 153

[219] Tweedie, M. C. K. (1957). Statistical properties of Inverse Gaussian distributions, *Annals of Mathematical Statistics*, 28:362–377, 696–705. DOI: 10.1214/aoms/1177706964. 153, 161

[220] Ueno, T., Hirano, S., et al. (2011). Stability of the Rayleigh distribution, *Fourth International Congress on Image and Signal Processing (CISP)*, 5:2376–78. DOI: 10.1109/cisp.2011.6100689. 250

[221] Usta, I. (2016). An innovative estimation method regarding Weibull parameters for wind energy applications, *Energy*, 106:301–314. DOI: 10.1016/j.energy.2016.03.068. 235

[222] Vanegas, L. H. and Paula, G. A. (2016). An extension of log-symmetric regression models: R code and applications, *Journal of Statistical Computation and Simulation*, 86:1709–1735. DOI: 10.1080/00949655.2015.1081689. 176

[223] Voinov, V., Nikulin, M., and Balakrishnan, N. (2013). *Chi-Squared Goodness of Fit Tests with Applications*, Academic Press, UK. DOI: 10.1016/c2011-0-05248-1.

[224] Wais, P. (2017). Two and three-parameter Weibull distribution in available wind power analysis, *Journal of Renewable Energy*, 103:15–29. DOI: 10.1016/j.renene.2016.10.041. 235

[225] Wald, A. (1947). *Sequential Analysis*, John Wiley, NY. 153, 160

[226] Wanke, P. and Leiva, V. (2015). Exploring the potential use of the Birnbaum–Saunders distribution in inventory management, *Mathematical Problems in Engineering*, 827246. DOI: 10.1155/2015/827246. 173

[227] Weibull, W. (1961). *Fatigue testing and analysis of results*, Pergamon Press. DOI: 10.1115/1.3640640. 241

[228] Wilks, S. S. (1932). Certain generalizations in the analysis of variance, *Biometrika*, 24:471–494. DOI: 10.1093/biomet/24.3-4.471. 233

[229] Wilson, E. B. and Hilferty, M. M. (1931). The distribution of chi-square, *Proc. of the National Academy of Sciences*, USA, 17:684–688. 204, 207

[230] Wishart, J. (1947). The cumulants of the z and of the logarithmic χ^2 and t distributions, *Biometrika*, 34:170–178. DOI: 10.2307/2332520. 233

[231] Wojnar, R. (2012). Rayleigh's distribution, Wigner's surmise and equation of the diffusion, *Proc. of the 6th Polish Symposium of Physics in Economy and Social Sciences*, Gdansk, Poland. 246, 251

[232] Wong, A. (2008). Approximating the F distribution via a general version of the modified signed log-likelihood ratio statistic, *Computational Statistics and Data Analysis*, 52:3902–3912. DOI: 10.1016/j.csda.2008.01.007. 231

[233] Woo, J., Ali, M. M., and Nadarajah, S. (2005). On the ratio $X/(X + Y)$ for Weibull and Levy distributions, *Journal of Korean Statistical Society*, 34:11–20. https://scienceon.kisti.re.kr/srch/selectPORSrchArticle.do?cn=JAKO200516610508191 241

[234] Yakovenko, V. M. and Silva, A. C. (2005). Two-class structure of income distribution in the: Exponential bulk and power-law tail, *Econophysics of Wealth Distributions*, Chakrabarti, B. K. (Eds.), pages 15–23, Springer. DOI: 10.1007/88-470-0389-X_2. 188

[235] Yoo, S. K., Cotton, S. L., Sofotasios, P. C., Matthaiou, M., Valkama, M., and Karagiannidis, G. K. (2017). The Fisher–Snedecor F distribution: A simple and accurate composite fading model, *IEEE Communication Letters*, 21(7):1661–1664. DOI: 10.1109/LCOMM.2017.2687438. 234

[236] Yoon, J., Kim, J., and Song, S. (2020). Comparison of parameter estimation methods for normal inverse Gaussian distribution, *Communications for Statistical Applications and Methods*, 27:97–108. DOI: 10.29220/csam.2020.27.1.097. 158

[237] Zacks, S. (1984). Estimating the shift to wear-out of systems having exponential Weibull life, *Operations Research*, 32:741–749. DOI: 10.1287/opre.32.3.741. 240

[238] Zar J. H. (1978). Approximations for the percentage points of the chi-squared distribution, *Applied Statistics*, 27:280–90. DOI: 10.2307/2347163. 207

Authors' Biographies

RAJAN CHATTAMVELLI

Rajan Chattamvelli is a professor in the School of Advanced Sciences at VIT University, Vellore, Tamil Nadu. He has published more than 22 research articles in international journals of repute and at various conferences. His research interests are in computational statistics, design of algorithms, parallel computing, data mining, machine learning, blockchain, combinatorics, and big data analytics. His prior assignments include Denver Public Health, Colorado; Metromail Corporation, Lincoln, Nebraska; Frederick University, Cyprus; Indian Institute of Management; Periyar Maniammai University, Thanjavur; and Presidency University, Bangalore.

RAMALINGAM SHANMUGAM

Ramalingam Shanmugam is a honorary professor in the school of Health Administration at Texas State University. He is the editor of the journals *Advances in Life Sciences*, *Global Journal of Research and Review*, *Journal of Obesity and Metabolism* (open access), and *International Journal of Research in Medical Sciences*, and book-review editor of the *Journal of Statistical Computation and Simulation*. He has published more than 200 research articles and 120 conference papers. His areas of research include theoretical and computational statistics, number theory, operations research, biostatistics, decision making, and epidemiology. His prior assignments include University of South Alabama, University of Colorado at Denver, Argonne National Labs, Indian Statistical Institute, and Mississippi State University. He is the president of the San Antonio chapter of the *American Statistical Association* and is a fellow of the International Statistical Institute.

Index

Printed in the United States
by Baker & Taylor Publisher Services